Girls and Mathematics:
From Primary
to Secondary Schooling

Bedford Way Papers   ISSN 0261—0078

1. 'Fifteen Thousand Hours': A Discussion
Barbara Tizard et al.
ISBN 0 85473 090 7

2. The Government in the Classroom
J Myron Atkin
ISBN 0 85473 089 3

3. Issues in Music Education
Charles Plummeridge et al.
ISBN 0 85473 105 9

4. No Minister: A Critique of the D E S
Paper 'The School Curriculum'
John White et al.
ISBN 0 85473 115 6

5. Publishing School Examination Results:
A Discussion
Ian Plewis et al.
ISBN 0 85473 116 4

6. Work and Women: A Review
Janet Holland
ISBN 0 85473 119 9

7. Science Teachers for Tomorrow's Schools
Arthur Jennings and Richard Ingle (eds.)
ISBN 0 85473 123 7

8. Girls and Mathematics: The Early Years
Rosie Walden and Valerie Walkerdine
ISBN 0 85473 124 5

9. Reorganisation of Secondary Education in
Manchester
Dudley Fiske
ISBN 0 85473 125 3

10. How Many Teachers? Issues of Policy,
Planning and Demography
Tessa Blackstone and Alan Crispin
ISBN 0 85473 133 4

11. The Language Monitors
Harold Rosen
ISBN 0 85473 134 2

12. Meeting Special Educational Needs: The
1981 Act and its Implications
John Welton et al.
ISBN 0 85473 136 9

13. Geography in Education Now
Norman Graves et al.
ISBN 0 85473 219 5

14. Art Education: Heritage and Prospect
Anthony Dyson et al.
ISBN 0 85473 149 0

15. Is Teaching a Profession?
Peter Gordon (ed.)
ISBN 0 85473 220 9

16. Teaching Political Literacy
Alex Porter (ed.)
ISBN 0 85473 154 7

17. Opening Moves: Study of Children's
Language Development
Margaret Meek (ed.)
ISBN 0 85473 161 X

18. Secondary School Examinations
Jo Mortimore and Peter Mortimore
ISBN 0 85473 167 9

19. Lessons Before Midnight:
Educating for Reason in Nuclear Matters
The Bishop of Salisbury, et al
ISBN 0 85473 189 X

20. Education plc?: Headteachers
and the New Training Initiative
Janet Maw et al
ISBN 0 85473 191 1

21. The Tightening Grip: Growth of Central
Control of the School Curriculum
Denis Lawton
ISBN 0 85473 201 2

22. The Quality Controllers: A Critique of
the White Paper 'Teaching Quality'
Frances Slater (ed.)
ISBN 0 85473 212 8

23. Education: Time for a New Act?
Richard Aldrich and Patricia Leighton
ISBN 0 85473 217 9

24. Girls and Mathematics: from Primary to
Secondary Schooling
Rosie Walden and Valerie Walkerdine
ISBN 0 85473 222 5

Girls and Mathematics
from Primary to Secondary Schooling

Rosie Walden and Valerie Walkerdine
with the assistance of Marilyn Hayward, Eileen Ward and Diana Watson

Bedford Way Papers 24
Institute of Education, University of London
distributed by Heinemann Educational Books Ltd.

First published in 1985 by the Institute of Education, University of London,
20 Bedford Way, London WC1H 0AL.

Distributed by Heinemann Educational Books Ltd., The Windmill Press,
Kingswood, Tadworth, Surrey KT20 6TG

Cover design by Herb Gillman

The opinions expressed in these papers are those of the authors and do not necessarily reflect those of the publisher.

ISBN   0   85473   222   5

British Library Cataloguing in Publication Data

Walden, Rosie
   Girls and mathematics: from primary secondary schooling. —
   (Bedford Way papers; 24)
   1. Mathematical ability   2. Women — Education   — Great Britain
   I. Title   II. Walkerdine, Valerie   III. Series
   510'.7'1041   QA135.5

ISBN 0-85473-222-5

Printed in Great Britain by Reprographic Services
Institute of Education, University of London.
Typesetting by Joan Rose

99/100/101-I1-001-114-0485

Contents

Preface

This paper is a *summary* of the work on the transition from primary to secondary school in the Girls and Mathematics Unit over the past three years. It condenses our arguments and the presentation of data, both of which are considerably more elaborated than this. More detailed and extended treatments of both theory and data will appear in later publications.

We should like to thank the staff and pupils of the schools in which we carried out fieldwork for their unfailing support and co-operation, and the local education authority for allowing us permission to undertake the study.

Both Eileen Ward and Marilyn Hayward were centrally involved in the work of the project: without them we should never have been able to accomplish as much as we have. Our thanks are also due to Diana Watson, who helped in the collection of data whilst on placement from the North East London Polytechnic, to Dr Patrick Slater, who developed the computer programmes used to analyse the repertory grid data and to Charlie Owen of the Thomas Coram Research Unit, who helped with the analysis and contributed to the discussion of significance testing in Chapter 3.

We remain, as ever, grateful to the Equal Opportunities Commission and to the Leverhulme Trust for providing the funding.

Finally, we wish to express special thanks to Professor Basil Bernstein for his support of our work and for time taken to clarify and extend our ideas.

R.W.   V.W.
November 1984

Rosie Walden was Research Officer in the Girls and Mathematics Unit, based in the Sociological Research Unit of the University of London Institute of Education and now teaches in Hackney, London E2.

Valerie Walkerdine is Director of the Girls and Mathematics Unit and Lecturer in Curriculum Studies at the University of London Institute of Education.

Chapter One
The Context

In *Girls and Mathematics: the Early Years* (Walden and Walkerdine, 1982), we explored some current explanations of girls' performance in mathematics. We noted that, although there was a prevailing view of girls' overall failure, this was both inaccurate and misleading. In the primary school at least, many girls were relatively successful. We examined how that success was understood and evaluated and suggested that further work was necessary to examine what happened to girls on moving from primary to secondary schooling. This paper reports such a study. Whereas we had suggested in our earlier study that girls' relatively poor performance at CSE and O level mathematics could be explained in terms of a *discontinuity* between earlier and later performance, the research reported here does not support such an explanation.

We suggested that although the phenomenon to be explained was girls' failure relative to boys to secure high grade GCE O level passes in mathematics and to enter post O level mathematical studies, the existing explanations operated as though there were consistent and continuous failure of all girls at all stages of their educational careers. We pointed to the fact that, up until the age of eleven (that is, the end of the primary school), girls were relatively successful in mathematics (APU, 1980/81/82; Ward, 1979) and that girls' performance in other subjects so outstripped that of boys that problems about criteria of selection to secondary school had been posed (Yates and Pidgeon, 1957). To that extent, then, it could be argued that explanations of performance which operated as though there were a simple and continuous poor performance were mistaken.

Our pilot study and review of previous research in the field (Eynard and Walkerdine, 1980; Walden and Walkerdine, 1982) argued that explanations which relied upon sex differences in spatial ability and sex-role stereotyping were insufficient to account for the phenomenon of early success, since both kinds of approach were premised on an expectation of failure. Our examination of the classroom performance of girls in the infant and nursery school revealed that the differences in play which had been proposed by some as an explanation (see Hartley, 1980 and Clarricoates, 1978, 1983) were neither statistically nor practically significant. Similarly, while there was evidence of sex-role stereotyping within the classrooms in which we observed this did not prevent some girls being in important, powerful and successful positions.

This led us to think it important to examine that period within girls' education in which it was presumed that the good performance found at eleven became the relatively poor performance exhibited at O level. With this in mind we proposed to examine the transfer of children from the top of the junior school to the first year of comprehensive schooling. In addition we proposed a similar set of observations of children in the fourth year of the comprehensive school, the time at which, within the school studied, the children were selected for O level or CSE entry.

We also argued for the necessity of carrying out a detailed examination of classroom practices in order to understand this phenomenon. Approaches using test data had revealed the discontinuity which we had highlighted but could not go very far towards understanding the production of discontinuity in performance. Observational work, relying as it does on the collection of data which cannot be easily subjected to statistical analysis, has been treated with some caution in the research world. However, the findings of this study made it absolutely essential to seek to understand the phenomenon in question.

We came to the conclusion that we would have completely to rethink the way in which the issue of girls' performance is to be considered. Even to treat the issue as one of 'discontinuity' implies that girls' performance simply deteriorates over time. As we shall explain later, we no longer feel the matter can adequately be explained in these terms. Instead, what emerges is a complex interplay between the way in which mathematical performance is understood, its classroom production, and gender differentiation in the assessment of good and poor performance.

In addition, the practical consequences of that differential assessment in terms of examinations entry, and subsequent attainment of CSE or O level passes, is very far from being a simple matter of a clear differentiation in ability leading to differential success. We shall argue that what is necessary is no less than a complete shift in the way that the problem of girls' performance is understood.

The historical context of girls and mathematics
In order to understand the necessity for a shift in the way in which the problem is to be envisaged it is necessary to outline in brief terms certain central concerns about the way in which the issue of mathematical education and the issue of girls and women has been understood. What we shall argue is that certain central conceptions about children, learning and mathematics have been crucial in providing the categories through which work is allocated and performance judged within both primary and secondary school classrooms. These conceptions, while claiming to be both universal and gender-neutral can, in fact, be shown to be both historically and gender specific (see, for example, Walkerdine, 1983a and 1983b).

We have argued elsewhere (Corran and Walkerdine, 1981) that modern conceptions of primary school mathematics are premised on certain ideas taken from cognitive developmental theory. These are, put simply, that mathematics is to be thought of not as 'facts to be stored', as in old conceptions relying on class teaching, rote memorization and rule-following, but in terms of individual development in which mathematical concepts are developed through a process of individual discovery. Class teaching gives way to individual or group work. Such approaches not only specify learning in terms of a progression of stages of cognitive development, but also set criteria for evaluating classroom performance in terms of the presumed acquisition or not of appropriate concepts. For our purposes what is important is the relation between classroom performance in terms of attainment and the judgement of acquisition of concepts. It is, in principle, possible to be successful within the mathematics curriculum but for that success not to be equated with the requisite conceptualization. In other words a mismatch is possible. It is this relationship, between classroom performance and what that performance is taken to be based on, which is central to the consideration of gender differentiation.

If we examine these assumptions about learning and teaching in the primary school in general and mathematics teaching in particular, certain distinctions between old and new mathematics emerge (Corran and Walkerdine, 1981). In the new mathematics conceptual understanding is taken to be the bedrock. Thus, to the old ways of 'chalk and talk', rote learning, simple and accurate performance, together with their allied practices — class textbooks, desks in rows, strictly bounded subject timetables, etc. — are contrasted the new, stressing individual development and the production of understanding. An accurate performance is in itself not sufficient; teachers will look for *evidence* that the success has been produced in the *correct* way. In development psychology such evidence is obtained by the administration of cognitive tasks. In the common sense of the classroom certain attributes which suggest such conceptualization tend to be taken as unofficial evidence that learning has occurred. Such attributes are activity, not passivity, 'flair', rule breaking, divergence, confidence, being articulate.

The teacher interviews which we conducted suggest that such characteristics are attributed in the main (though certainly not exclusively) to *boys*. There is, thus, a clear relation between proper learning and proper masculinity. Proper femininity, stressing passivity, tends to mean that any success is interpreted as being produced by the wrong means. In this way, we can understand how the practices in which the teachers engage produce the categories for their evaluation of performance. It is not so much that they show 'bias' or simply 'mislabel', but that the characteristics displayed by girls and boys in the classrooms lead teachers to read their performance in different ways and therefore to do different things about it. In this view, the girls' success is not really success at all. Clearly we need to monitor just how and why girls behave in this way in the classrooms. Yet is is not a simple matter of teacher bias against femininity. Rather is it a complex relation between the theory and practice of learning and teaching, leading to gender-differentiated production of success and failure. As we shall see some girls *do* succeed in producing good performance in 'the correct way' and it has certain consequences for them. But the answer, as we shall argue, is not a simple one of 'making girls more like boys' in order to make them successful.

Let us take this issue further, for it relates not only to what counts as proper learning but also to the characteristics of the proper learner. Clearly if concepts are taken to be acquired through discovery then the ideal learner is, to quote the Plowden Report (Central Advisory Council on Education, 1967), 'naturally inquisitive'. 'The basis of all educational questions', the Plowden Report tells us, 'is the nature of the child himself.' The ideal child is by nature enquiring, discovering, creative, playful. It is a view of human nature founded in a set of assumptions about a natural sequence of child development. However, further analysis reveals that the nature taken to be universal and gender-neutral is in fact ascribed to the ideal boy:

> The high-spirited mischievous child is traditionally regarded with affectionate tolerance. 'Boys will be boys' . . . a boy who never gets up to mischief, it is suggested, is not a proper boy . . . Yet boys need a sense of adventure.

Since girls rarely conform to this ideal, they cannot therefore meet the specifications of 'the ideal child'. If proper understanding is taken, both within the practice of mathematics teaching and the theory on which it rests, to be the result of a set of activities in which proper 'high spirited and mischievous boys engage, by their very nature', then it might be assumed that there is a problem for girls. We shall argue that to understand the performance of girls we have to see how it is produced and evaluated. In the main it is girls who are taken to produce good performance but for the wrong reasons. They are not high-spirited but responsible and hard-working. But, since play and adventure are taken as those characteristics which describe the proper learner, it is not surprising that teachers' evaluation of girls often suggests that they are not really clever, not necessarily because of poor performance *per se,* but because they do not engage in the right sort of classroom activities. Put at its most simple: girls obey and follow the rules; rule-following is regarded as the wrong kind of learning; so girls are put in a double-bind. The rules concerned are both behavioural and organizational rules in the classroom and those of mathematical knowledge (Corran and Walkerdine, 1981). But it is precisely the breaking of these rules — by being naughty and mischievous on the one hand, and challenging the authority of the teacher's claim to the knowledge on the other — which is taken as evidence of real understanding, often quite outside any consideration of whether actual attainment is very good. What can be observed, then, is the acting out in practice of the theoretical double-bind created within the theory of mathematical performance itself, in such a way as paradoxically to set up rules which are to be broken and to punish those who obey and follow the rules for doing so.

Some previous approaches, in acknowledging girls' rule-following and good behaviour, have suggested rather different explanations and courses of action. Thus, Shuard (1981, 1983) has argued that young girls' success in those kinds of mathematics taken to require only rule-following rather than higher-order conceptualization is the result of practices of teaching which, contrary to expected good practice of allowing enquiry, foster rule-following by stressing only lower-order activities. This fails to explain, however, why boys in such classrooms do not also end up performing well in such tasks. In other words, emphasis on the quality of teaching does not, in and of itself, explain gender-differentiated behaviour. A second line of explanation, using notions derived from the 'hidden curriculum' (for example, Clarricoates, 1983; Whyte, 1983), would argue that girls are expected to conform to a female stereotype which is passive and dependent, so that they are not rule-breaking and mischievous like boys.

The solution, then, is taken to be to encourage girls to be more inquisitive, naughty, and so on. We wish to raise two points of contention here. Firstly, such an approach understands girls' different behaviour as the result of 'hidden' processes, not apparent in the overt curriculum. Secondly, teachers are blamed for their imposition of stereotyped attitudes which if removed would allow a change in girls' behaviour, allowing it to be more like that of boys.

Our approach is critical of such an interpretation for several reasons. Firstly, our argument stresses the importance for understanding the production and evaluation of girls' performance in mathematics of the theory and practice underlying mathematics teaching. We make certain assumptions about the relationship between performance and human nature, such that there is no simple and direct problem of poor performance but rather a complex relation between the production of performance, attainment and the evaluation of that attainment in terms of what is taken to be its cause. It is not the case that girls do not achieve and that they would do so if teachers allowed them to be more like boys, but that when they do achieve it is often taken to be for the wrong reasons. What we have here is certainly not the result of reactionary, old fashioned or stereotyped values on the part of the teacher. Indeed, quite the reverse is often the case. Such evaluations in performance are the result of adherence to that kind of practice which is taken to be modern and gender neutral and is, therefore, adopted by those teachers considered to be most progressive. They, too, should be understood as caught up in a complex web in which their attempts at helping girls result in practices which appear to help maintain their exclusion from higher-level attainment, as we shall discuss with respect to examination entry at sixteen.

What is evident is that certain conditions of good practice, such as the development of individual work, allowing children to go at their own pace, and so on, form the criteria which teachers recognize and can talk about easily. However, they also specify standards of behaviour and presentation which are at once to be followed and to be broken, since proper performance depends on their being broken.

In an earlier report (Corran and Walkerdine, 1981) we were critical of the distinction between propositional and procedural knowledge in the mathematics curriculum. We argued that children recognized and operated within procedural rules in order to carry out those tasks the successful completion of which was taken as evidence of propositional knowledge. The distinction between propositions as 'real understanding' and procedures as 'rule following' characterizes the gender-differentiated evaluations of success. Challenging and breaking procedural rules applies equally to classroom behaviour and the procedures for operating within mathematical knowledge itself. We shall demonstrate the way in which the challenging of mathematical rules provides a challenge to the authority of the teacher and a power struggle which is antithetical to femininity. Those children described as having 'real understanding', 'brilliance', 'flair', etc., can be shown to engage in certain practices within the classroom which relate to the challenge of rules and of the authority of the teacher-custodian of knowledge. The failure of many girls to display such performance, while at the same time producing high attainment (classed as 'hard work') puts them in a double-bind, as we have said. However, this failure to challenge cannot readily be squared with the procedural-propositional distinction, in which 'real understanding' is the result not of positionings within classroom practices, but an individualized sequence of cognitive development. Cognitive development in this model is a quasi-natural phenomenon, which is formed through the production of logico-mathematical structures, not social relations and practices. In challenging the assumptions about the performance of girls we are thereby questioning the theoretical basis on which that performance is understood (see also Henriques et al, 1984).

Related to this, we can also explore why, for girls, innovation and challenge is easier in English than mathematics. Basically, while both kinds of knowledge make challenge and imaginative work possible, the threat to the authority of the teacher in mathematics classrooms is more basic and profound than in English lessons. A demonstration of creativity in English is relatively private, more open-ended than it is in mathematics, where certainly 'elegance' of proofs is to be applauded, but the significance of 'knowledge' and 'truth' control is very different from English literature. Since in mathematics open challenging is difficult for girls, differentiated performance in the two subjects is hardly surprising.

Girls attain power through the regulation of the rules of the classroom (relating to teacher identification, discussed elsewhere, e.g. Walden and Walkerdine, 1982). For girls to break such rules, to stand out, to be naughty, has very different consequences for them than for boys. The gender-differentiated marking of challenge, naughtiness and rule-breaking is crucial. We would argue that concentration on the hidden curriculum ignores vital processes in the regulation of knowledge and of educational practices, and that examination of such regulation is extremely important in understanding the position of girls.

It is not, then, simply a case of making girls more like boys. Rather is it an issue of questioning the basis on which judgements of performance are made, which on one level leads to the questioning of theories of cognitive development. (Walkerdine, 1983b; Gilligan, 1982/83 for example) and on another led us to question the assumption of a discontinuity in performance. The difference in examination entry and success at sixteen turns out not to be a simple matter of a decline in performance but indeed the result of a continuity in the sort of judgements made, which culminated in two practices in the comprehensive school in which our observations were made. The first was that since some girls' performance was taken to be the result of hard work, lack of confidence and rule-following, they were not pushed. In this instance allowing them to go at their own pace meant, in a word, slow. They were not pushed; they were protected. A consequence of this protection was that girls were entered for CSE rather than O level because the former (in its Mode Three form) was school based and therefore considered less stressful and more likely to result in success.

Secondary school mathematics took up the 'new mathematics' rather later than primary schools. But nevertheless, the same criteria hold: good practice is taken to be individual rather than class-based, and so on. The scheme in operation in the school in which we worked was called SMILE (Secondary Mathematics Individualized Learning Experiment). The acronym SMILE testifies to the attempt to produce better results through a more free and therefore happier practice, children self-motivated by going at their own pace.

Schemes with the individual approach of SMILE, used in the secondary school we investigated, avoid explicit competition and open combat. Although it is taken as based on 'real understanding', its individualized approach permits children to proceed quietly and without open challenge or competition. We shall demonstrate that many girls like SMILE precisely because it makes them feel safe, whereas often boys dislike it, finding it babyish. So the scheme designed as progressive and helpful to less confident pupils actually appears to help regulate and maintain such lack of confidence.

In addition to this, the development of Mode Three CSE examinations in the 1960s permitted a shift from what many saw as the tyranny of external evaluation. Mode Three examinations were school based and relied heavily on continuous assessment as well as set examination. They were, therefore, seen as more in keeping with modern ideas about learning and pedagogy (Barnes, Britton and Rosen, 1969; Postman and Weingartner, 1981; Smyth and Stephenson 1971; Young, 1970).

Many enlightened teachers, therefore, were much more favourably disposed to Mode Three CSE examinations than O levels and, given that a

CSE Grade One was taken to be the equivalent to an O level pass, the teachers believed an equity between the two systems was possible. Consequently, it became the practice to put less confident children in for the Mode Three examinations, and many of the less confident children were girls.

What has to be explained are the processes within the classrooms themselves which lead to girls' apparent lack of confidence, to the gender differentiated examination entry and to why when entered most girls do not achieve Grade One passes. We shall argue that such processes and practices are potentially explicable and that we can begin to demonstrate how such differentiation is achieved. Girls' performance can then be understood as intimately bound up with the criteria for its production and evaluation. In this sense performance differences at sixteen relate both to conceptions of mathematics, theories of development and learning, and to the practices in the classroom. It is this complex interplay of conditions which we aim to demonstrate in the analysis of classroom data which follows.

Chapter Two
Data Collection

In this chapter we shall describe the methods we adopted to pursue our study. Our concern was to examine in more detail those processes which became apparent from pilot work in primary schools and to investigate our hypothesis of discontinuity of performance from primary to secondary school. We needed to take account of the kind of data already collected in other work and also to be able to move beyond explanations offered by existing approaches. In this respect it was crucial to devise an investigation which mixed quantitative and qualitative data. Existing approaches took as their basis either quantitative performance data or qualitative accounts based on a theory of stereotyping and the hidden curriculum, both of which tend to be better at explaining failure than success. Mixing the two types of approach therefore seemed a more profitable approach.

The sample
In our pilot work (Walden and Walkerdine, 1982) we chose our sample on the basis of the teachers' choices of those girls, and boys, whom they felt were good and poor at those activities which were understood as mathematics in the nursery and infant school (see also Corran and Walkerdine, 1981). In this way we were able to understand the relationship of behaviour and performance in the context of the classroom.

We adopted a similar method of selecting a sample in this study. We began by asking teachers in two primary schools (J1, J2) which were feeder schools for one comprehensive school (S1) to choose for us children they considered good and poor at mathematics. This sample, originally fourteen, eleven girls and three boys (twelve in the first year of secondary school, a boy and a girl having gone to different schools), was observed in their last year of primary school and their first year of secondary school (aged 11 and 12), a period covering a crucial stage of transition in their educational careers embodying changes in size and structure of school as well as organization, teaching methods, subjects taught, and so on. Ideally, we should have liked to follow these children through to their fifth year, monitoring their progress and development throughout the important phase of their secondary years, but this was obviously unrealistic. So we chose to look at a very small sample of children in their fourth year at the secondary school (aged 14 and 15).

We chose the fourth year as the important preparatory stage for the public examinations and took a sample of children in their fourth year just prior to their selection for the public examinations to look at how choices were made on whom to enter for either CSE or O level (see Chapter 1). Although not a perfect method, it did enable us finally to collect a cross-sectional set of data about girls' development from the end of primary school to just prior to leaving school and to see how the girls understood the tasks they were set and how they were themselves produced as learners of mathematics.

Analyses of classroom practices
Our study uses a deliberately small sample: that is, it was conducted in two junior schools and one comprehensive. However, the word 'small' here is deceptive since we have collected many hours of audio and video recording, which in the comprehensive school alone, represents analyses of the practices of five first year classrooms.

The main point that we would want to make concerns validity. It is common, with analyses of quantitative data, to assume validity is given on the basis of adequate statistical criteria. This means that the validity of the underlying assumptions and theoretical considerations generating the research does not come into question. Yet, of course, these matters are central to any understanding of what might validly be claimed on the basis of empirical work. While it is quite common, in the tradition of work on mathematics, to put down as problematic work requiring interpretation, we aim to demonstrate that current work relying on statistics also poses real problems of interpretation which largely go unrecognized. In that sense, then, we shall argue that there is no empirical work which does not raise problems of interpretation. It is important to make assumptions explicit and particularly one's method for generating, reading and interpreting data. We take this to be fundamental to work of any kind. Criteria for determining validity cannot be restricted to statistical means.

One major factor concerns the issue of concentrating on one or two schools. It is quite common to select schools in a sample by holding constant certain factors, for example, social class of intake. Yet each school, when examined in detail, presents a highly complex institution which deserves examination in its own right and in its specificity. Our analysis takes therefore as a basic premise the assumption that school and classroom practices develop in certain historical conditions of possibility. For example, we shall argue that this comprehensive school, favouring as it did the entry of pupils for Mode 3 CSE rather than O level, produced specific effects for the attainment of certain girls. Clearly these precise effects would not be demonstrable in a school with a different policy. Attempting to understand therefore how attainment is produced in *both* general and specific ways, necessitates a mode of data collection and analysis which can engage with both the general and the specific, and not ignore one at the cost of the other. Although we are only just beginning to examine such an approach to fieldwork we suggest that the aim for validity is not helped by a reduction of complexity and specificity if precisely those factors play an important role in producing the phenomena under study. It is clearly inappropriate to treat *girls* and *boys* as unitary and distinct categories, rather than examine the production of gendered attainment in its specific locations. It is such an approach which has tended to produce a set of fixed and immutable sex differences, or basic characteristics, which are then reified and built into the methodology as taken-for-granted categories.

In that respect our analyses of classroom practices, while they fall within a broadly ethnographic or interpretative tradition, attempt to utilize insights from post-structuralism (see Henriques et al., 1984). In particular we are interested in relating the current 'truth' of the production of mathematics attainment, to how this becomes part of teachers' and pupils' common sense and therefore how the participants' behaviour itself is produced and understood in a way which so powerfully affects how the pupils fare within the school system.

In that sense, while our work draws upon previous ethnographic traditions, and feminist work on classrooms and the hidden curriculum, it differs from them in certain crucial respects. Specifically, one focus of our study is the effects of the *overt* categories utilized within the mathematics curriculum. We argue that these actively help to produce the phenomena they claim to be describing. In that respect our framework is very critical of how such ethnographic or interpretative traditions are ignored in quantitative work in this field, work often built upon unquestioned assumptions about the phenomena under interrogation. In taking apart the conditions for the production of the present 'truth' of gendered mathematical performance, we can begin to examine the way in which practices themselves help to generate differences in attainment within certain specific historical conditions of possibility. In that sense, then, such work challenges claims which might be made about trans-historical studies of sex differences, and of 'human nature', not by reference to a nature-nurture opposition but by examining how the 'truth' of current claims is itself produced, reproduced and lived in the positions created in the practices themselves. In this analysis the power of such knowledge is central, for it has real effects in determining the life chances of children through its claims to tell the truth about their attainment (Foucault, 1979).

Problems
Our major problems concerned areas which it would have been difficult to foresee. Our fieldwork schedule, which was arranged by schools some eight months in advance, was affected by a variety of 'accidents': illness, absence, timing of school journeys, epidemics of scarlet fever, maternity leave, early retirement, promotion, amalgamation of schools — none of these could have been foreseen, and they had to be coped with as they occurred. Overall, we were extremely fortunate in the high level of co-operation which we received from the teachers and the children involved. Since our method involved spending a long time (a term) in any one classroom it was possible to build up a useful rapport with the subjects of the research. Familiarity with the researcher and equipment is extremely important to our method.

The problems that we had arose mainly in the secondary school. The twelve children left in the sample were spread between five different tutor sets in a ten-form-entry first year. Effectively we were involved with nearly a hundred children. Mathematics for the first year was timetabled so that at least four first-year sets at any one time were being taught. The tightness of the schedule and the need to fit in a period of observation prior to any recording and interviewing meant that any absence or illness on the part of either researcher or child or teacher needed to be compensated for at another time in the schedule.

Once the data was collected, however, the first stage of analysis was to transcribe all videotapes and all interview tapes. It is very difficult to decide in advance how long this will take: much depends on how much is said, how much is audible, and so on. On average a videotape takes up to two days to transcribe in its entirety and we had 50 in all from the secondary school and 28 from the primary school, plus 27 audio tapes including 74 repertory grid interviews (30 fourth-year junior: 32 first-year secondary: 12 fourth-year secondary) and a total of 14 teacher interviews (throughout the three year groups), totalling 84 tapes in all. Overall, though, the methods used, time-consuming as they were, provided us with the sort of data which we had expected — rich and detailed.

Chapter Three
Test Data

Most of the data used to explain sex differences in mathematical performance take the form of mathematical tests. Briefly, explanations of differential performance have centred on a particular set of assumptions which can be summarized as follows:

1. Boys are significantly better on test items requiring 'spatial ability';

2. Boys are better at more complex items requiring abstract thought, problem solving, breaking set, conceptualization;

3. Girls are better at simple, repetitive tasks requiring low-level skills such as rule-following (see Shuard, 1981; APU Primary and Secondary Surveys, 1980a, 1980b, 1981a, 1981b, 1982a, 1982b; Sheffield City Polytechnic, 1983).

The reification of categories 'girl' and 'boy' help to produce explanations which favour sex-specific characteristics. This means that more complex analyses of masculinity and femininity are not possible. In addition there are certain important implications of current work using test and survey data which have been poorly treated in the literature. These relate to the, often hidden, problems of interpretation and, therefore, of determining the validity of findings using statistical criteria. Such findings are often treated as 'hard' evidence, as compared to the 'soft' data of the more interpretative traditions in psychology and social sciences. As questions of truth and validity are central to our argument, we shall introduce this chapter by a review of some problems with the interpretation of existing test data relating to sex differences in mathematical performance (extending arguments we have used previously: Walkerdine, Walden and Owen, 1982).

Interpretation of performance data

The issue of the interpretation of performance data relating to girls is important for several reasons. The search for difference in performance has had a long and infamous history, in which claims about the 'reality' of differences have been produced in support of arguments about the position and status of women: in other words, the political consequences of such statements about girls and women are profound. One of the important issues about research on sex and gender has been the overwhelming use of methods which search for *differences,* usually differences which can be quantified and stated to be real by the use of statistical techniques. This has certain important consequences. Firstly, that similarities are usually treated in terms of their failure to show significant differences: in other words, similarities become a non-result. Now, while we do not want to get stuck on the important theoretical issue of gender similarities and differences, what we would point out here is the importance of the way that questions are proved and techniques used to produce certain kinds of data on which only certain kinds of claims can be made, while other issues are ignored. We would concur with statements made by Tessa Blackstone and Mary Warnock (1982) in *The Times Educational Supplement* about the problems of continued emphasis on girls' failure and lack of confidence. Nothing succeeds like success but, similarly, constant emphasis on girls' lack of attainment means that their successes tend to go unnoticed.

This brings us to the point at issue, the interpretation of statistical data. In our earlier Bedford Way Paper (Walden and Walkerdine, 1982) we quoted from the Assessment of Performance Unit surveys, pointing out that they emphasize the trivial nature of any differences in mathematical performance at age eleven. However, various commentators, remarking on these statements, have taken the view that we have misinterpreted the survey data by claiming the triviality of the differences because some differences are statistically significant. For this reason we feel that it is crucial to explore the issue of statistical significance with respect to performance data on girls and boys. If significance is used as a basis for claiming that girls relative to boys are performing poorly in mathematics in the primary school, then this has to be investigated seriously.

An advantage of large-scale sample surveys is that they allow a fair degree of precision in describing the population in question. Unfortunately, the routine application in these situations of the significance testing procedures developed for small-scale experiments can be misleading. There is a close link between sample size and statistical power, that is, the ability of a statistical test to detect a difference. Consequently, in large surveys trivially small differences may be highly statistically significant, and this significance may be deceptive. Over-evaluation of small differences may be seen in discussions of some recent surveys of primary school mathematics attainment: some commentators have made much of very small differences in performance between girls and boys.

The three studies on which we shall concentrate are the first two Assessment of Performance Unit primary mathematics surveys, for 1978 (APU, 1980a) and 1979 (APU, 1981a) and the Schools Council study (Ward, 1979). The first APU study involved about 13,000 11-year-old pupils, each question being answered by about 1,500 pupils. The second survey involved about 14,500 pupils, with each question answered by 1,000 pupils. Ward (1979) tested 2,300 10-year-olds, each question being answered by about a quarter of the pupils. All three studies are very circumspect in their conclusions with regard to sex differences. The first APU study states: 'The data on sex differences show a slight, and generally non-significant, advantage to the boys in most sub-categories, but girls perform significantly better statistically in computation (whole numbers and decimals)' (p.72); and 'The boys' mean score is significantly higher statistically in three sub-categories: length, area, volume and capacity, applications of number, and rate and ratio' (p.68). Note the emphasis on *statistically* significant differences. Earlier, the report states: 'Throughout this report references are made to statistically significant results but the educational implications of these are usually left to the reader' (p.11), but it goes on to caution that: 'If very large samples have been used even small differences can become statistically significant' (ibid.).

The second APU study is equally unemphatic, saying only that: 'the boy's mean score is significantly higher in five of the sub-categories in 1979: the two measure ones, concepts (fractions and decimals), applications of number, and rate and ratio' (p.60); and 'the boys generally obtained higher mean scores than the girls although the sex difference is only statistically significant in five of the sub-categories' (p.64). This report includes an appendix on statistical significance, again emphasizing that 'statistical significance does not provide an indication of the educational significance of a difference . . .' (p.96), but failing to mention the importance of sample size.

Ward (1979) says this about sex differences in his study: 'The girls in the sample did slightly better at the straightforward computation questions than the boys. The boys made up for this by performing slightly better on problems put in words and those involving an understanding of the structure of number' (p.39). He does not use the term 'significance' nor does he discuss significance testing.

So all three give little emphasis to sex differences, and where statistical significance is mentioned, caution is stressed in its interpretation. Yet this caution is not always observed, so that Howson (1982) points out that between the first and second APU studies 'the boys had notched up significantly better performance in two further sub-categories', with no indication of the magnitude of these differences. In fact these two further sub-categories had differences of 0.8 units on each of 'money, time weight and temperature' and 'concepts (fractions and decimals)'. (The previous year these differences had both been 0.5 units.) The standard deviations for these sub-categories are 7.2 and 8.3 respectively, so that even the larger difference is only equivalent to about one and three-quarter IQ points — a trivially small difference by any reckoning.

The largest sex difference in the second survey was of 1.4 units in the sub-category 'applications of number'; with a standard deviation of 7.5 this is approximately equivalent to three IQ points. (Whilst not wishing to overemphasize the value of IQ tests, nor to imply they are unproblematic, just for comparative purposes here are some reported IQ differences: the average IQ difference of identical twins reared apart across five studies was 7.2 points (Johnson, 1963); the advantage of being born in June over being born in January was once reported as 2.2 points (Pintner and Farlano, 1933).)

How could such small differences be considered important? Part of the problem is the seductive word 'significant': surely something that is 'significant' (with or without the adjective 'statistically') must be important. This concept of 'significance' is part of the pidgin statistics of social science; researchers are pleased when they 'find' it and journal editors use it as the necessary stamp of worthwhile research, and yet its meaning is not sufficiently appreciated. Now and again there is a complaint against the abuse of significance testing (e.g. Morrison and Henkel, 1970; Atkins and Jarrett, 1979) but things usually go on much the same. Conventional significance tests test a null hypothesis that in some ideal population there is exactly zero difference in the mean or some variable in some chosen groups. In practice no one believes this strict null hypothesis to be true, and the question is whether the differences which do exist are big enough to take into consideration. Typically researchers carry out an experiment or make observations, and then carry out a test of significance. The size of the groups is usually determined by factors such as time and money available for the project, how easy it is to get people to take part, and a feel for what is generally acceptable. Rarely if ever is any consideration given to the power of a test.

The power of a statistical test is the probability that it will yield statistically significant results at a given level of significance when a difference of a certain size does exist. Consider the case where the true difference between two populations is half a standard deviation (about eight points of IQ). Then, even with two samples of thirty, the chances of getting a difference 'significant at the 5 per cent level' is less than a half — that is, half of such experiments will *fail* to detect this real difference. This is the power of the test. To achieve a power of three-quarters would require samples of about sixty. This is getting quite large for a psychological experiment, and half a standard deviation is quite a large effect, so the implication is that social science experiments are not very powerful. No wonder we are pleased when we get 'significance'!

As the sample size increases the probability of detecting a given difference (the power) also increases; alternatively, increasingly small differences will have an increasingly high probability of being statistically significant. As Nunally (1960) put it, 'If the null hypothesis is not rejected, it is usually because the N is too small. If enough data are gathered, the hypothesis will generally be rejected' (p.643). With two samples of 100 the power, for a difference of half a standard deviation and the 5 per cent significance level, is 0.94. Sample sizes of 750 give this same power for differences of less than 0.2 standard deviations, and a power of 0.5 for a difference as small as 0.1 standard deviations (Cohen, 1977). So samples of the size used by the APU mean that small differences between the scores of girls and boys are highly likely to be *statistically* significant, even though they may be of no educational significance.

Shuard (1981, 1982) draws on data from all three of these studies in two discussions of sex differences in primary school mathematics attainment. In an article in *The Times Educational Supplement* (1981) she focuses mainly on Ward (1979), but also draws on the APU studies. In this one-page review she uses the words 'significant' or 'significantly' fourteen times; for example, 'Out of 91 questions used in the tests, there were eleven on which girls performed significantly better than boys (at a 5 per cent level of significance or higher) and fourteen on which boys performed significantly better than girls.' It is unlikely that most readers of *The TES* would know what a '5 per cent level' means, but they would know what 'significance' is.

Shuard gives some detail on these significantly different questions, but ignores the ones that are not significant, even though the content of these items is sometimes almost identical. She says that Ward (1979) 'gives a picture of the 1974 crop of 10-year-old girls and boys setting out towards secondary school mathematics with some of their strengths and weaknesses already fixed', but she does not mention that the APU studies show that regional differences are generally far greater than sex differences. She does suggest that strengths and weaknesses of the Welsh sample very much reflect those of the girls, and suggests that this may reflect an overemphasis on computational skills in both cases, but she never mentions the relative size of sex and regional differences: she is blinkered by the 'significance' of the results.

Shuard, in her 1982 paper, again draws on the Schools Council and APU studies, but also includes a lot of other work. Discussing non-mathematical tests she rightly points out that 'on all tests, however, the overlap between the sexes is very large, and it would be a gross distortion to expect that most boys would be better than most girls' (p.279) on any particular task. Yet when it comes to mathematics she is quite willing to talk about the differences between girls and boys as if they were two quite distinct groups. In addition to this, in her *TES* article (1981) Shuard clearly has in mind a view of the issue of girls' performance in mathematics which is the driving force behind her analysis of the Schools Council data. This is that girls perform better than boys at computation and those aspects of mathematics considered as low-level, while boys' performance is better on more complex aspects of mathematics, particularly 'spatial' questions.

There are several points of contention here. Firstly, Shuard writes as though girls were only good at computation, whereas the results reveal, as we have argued, that on the majority of items there are no sex differences in performance. In other words, it is a gross misrepresentation of the data to suggest that girls are only good at computation. It is, however, a necessary slippage if it is to be argued that girls' performance is not up to standard. Shuard lays the blame at the door of primary schools which, she claims, stress low-level computation because it is easy to teach, thus hindering the development of 'real' mathematics. So, girls are not 'only' good at computation. But we can see that the way in which the argument is posed actually fully negates the performance of girls by implying that even in those areas where they are successful, their success is only low-level and can therefore be discounted.

This brings us to the second point of contention which is the distinction between computation as low-level and spatial tasks as 'real' mathematics. There are good reasons for Shuard to make this judgement, which are backed up by the data on sex differences in certain spatial tasks, though even here the issue is far from clear-cut (see Walden and Walkerdine, 1982). What is particularly contentious, however, is the down grading of certain aspects of mathematics as not 'real', or as only 'rule-following', and the upgrading of others. It so happens, of course, that since girls do well at the 'low-level' ones, there is a failure to point out that they also perform well on other aspects considered 'real'! The debate about the importance of rule-following in mathematics is far from closed (see Howson, 1982; Corran and Walkerdine, 1981; Walkerdine, 1982). It is a moot point whether rule following should be considered low-level at all or whether 'real understanding' plays the part in school mathematics which some mathematics educators would have us believe.

If we continue this examination by looking at the APU secondary surveys (APU, 1980b, 1981b, 1982b) similar points can be stressed. Although there are larger sex differences than in the primary surveys, the results need to be treated with the same caution. The sex differences in mathematics attainment found in the primary surveys were very small. In the three secondary surveys the differences are both larger and more often statistically significant. They are also consistently in favour of boys: 'In all three surveys, the mean scores of the boys have been higher than those of the girls in every sub-category, with only one exception . . . These differences have been statistically significant in up to eleven of the fifteen sub-categories of content' (APU 1982b, p.143). None of the three secondary surveys give standard deviations for the sub-category scores, so it is not possible to make simple direct comparisons with the primary surveys, nor to represent the size of the sex differences as notional 'IQ point differences'. However, it is possible to compare the size of the sex differences with the sizes of differences on the other 'background' variables.

The differences between boys and girls ranged from one or two per cent to about eight per cent. They were considerably smaller than the differences between pupils living in metropolitan versus non-metropolitan areas, and were totally swamped by differences between the regions of the United Kingdom or differences between schools having high or low percentages of free school meals. So the differences are still small, both in absolute and relative terms.

Two of the surveys (APU, 1980b, 1982b) discuss not just the means for girls and boys, but also the distributions. For both surveys, 'the higher mean scores of boys over girls are largely due to the greater preponderance of boys among the high scorers rather than girls among the low scorers' (APU, 1982b, p.73). For example in the first survey boys formed 50.5

per cent of the total sample, but were 58 per cent of the top 25 per cent of pupils, and 61.5 per cent of the top 10 per cent.

The third survey looked for consistencies between the three secondary surveys and for continuities between the primary and secondary surveys. For consistency they looked at the pattern of sex differences in the sub-categories, and found an average correlation of over 0.6 between the three surveys. It is suggested that the areas where boys' mean scores are largest relative to girls are those of mathematics which 'are important in several secondary curriculum options, such as physical sciences, technical drawing and woodwork which are taken by more boys than girls' (ibid. p.143). This is discussed more fully in the report.

There is no suggestion that the mean differences between girls and boys found in the primary surveys are consolidated or amplified in the secondary years, i.e. that the earlier smaller sex difference is just an early indication of what is to come later. However, it is suggested that the 'profiles of relative strengths and weaknesses across the different areas of mathematics' (APU, 1982b, p.145) *are* consistent. So that the things the girls are relatively better at in the primary years they stay relatively better at (e.g. girls are better at computation than at measure), *but* 'the boys' profile of attainment moves up about two percentage points relative to the girls' (ibid.). Consequently, 'Whatever factors are causing the differences in the profiles of attainment are already operating by the time the pupils are in their primary schools' (ibid.).

It is important, then, to recognize that we are talking of sex differences which, where they are found, are no greater than four IQ points (for example, scores of 112 and 116). What is important with such an example is not in any sense a truth or validity which stands outside the practices in which such differences have real and practical effects. For example, in so far as such scores will be utilized as cut-off points for educational provision, and in as far as such scores will be used to determine the opportunity and experience and curricula given to girls and boys, then we can say that they have 'real effects.' These effects have a materiality and a power which stands outside any guarantees of truth or interpretation in a 'basic' sense. We might well conclude that nothing can be usefully said about differences between scores of 112 and 116. What matters are the consequences for practice. It is these that we shall argue in the course of this study are of crucial significance in the regulation and operation of school mathematics practices themselves. Such an analysis of the power of these 'truths' is crucial because it allows us to examine the complexity of the production of the 'truth' about girls' performance and the effects on the schooling of girls.

Primary mathematics test
In both primary schools in which we observed classes were given a mathematics test. The test, which we devised in conjunction with the teachers, was to enable us to have some measure of performance against which to compare the teachers' judgements.

The test was similar to the comparability test (used at 11-plus in inner London schools — see Walden and Walkerdine, 1982 for more detail) and drew on the Assessment of Performance Unit's five categories of mathematics (see first and second primary surveys APU, 1980a and 1981a). These were geometry, measures, number, algebra, and probability and statistics. Neither test was timed since, as the APU puts it (1980a para.3.6): '. . . pupils would be more easily able to demonstrate what they knew if the test was short and they were not required to work against the clock'. Both concepts and skills were tested. By the APU's definition (APU 1980a, paras. 1.33 and 1.34), a concept 'involves the recognition of a relationship', whereas 'skills are learned routines'. Thus we hoped to obtain, on a much smaller sample, results which could be looked at in the light of the APU's analysis. It has been suggested by such writers as Shuard (1981, 1983) and subsequently the Sheffield Polytechnic team (1983) that there are sex differences in responses to various *types* of questions, i.e. girls are better at computation (rule-following) whereas boys are better at geometric (spatial) questions, so we included similar types of questions to those mentioned. Our test consisted of twenty-three questions, with a possibility of a maximum score of 40, and was administered to all sixty-six children in the top junior classes of the two schools as part of their normal mathematics lessons, with no time limit on completion.

Each question was coded as to whether it had been answered rightly, wrongly (attempted and wrong), or left unanswered (not attempted at all). The results were then analysed by school, by sex, then by school and sex according to how the question had been answered. The results were as follows.

Type of response
When all the results were analysed it was found that there were no sex differences in the numbers of questions answered correctly, answered wrongly, or left unanswered. The means for boys and girls on the three categories were as follows:

Table 1: Sex: type of response

	N	Right	Wrong	Unanswered
Girls	37	23.8	11.4	4.8
Boys	29	23.8	11.3	4.9

However, if we look at the same analysis by question answered but broken down between schools we find interesting differences:

Table 2: School: type of response

	N	Right	Wrong	Unanswered
J1	30	23.8	9.6	6.6
J2	36	23.8	12.8	3.5

Whilst the number of questions answered correctly remains the same between the schools as it had for the sexes, the interesting variations occur between the other responses. School J2 has nearly a quarter more wrongly answered questions than school J1. J1 on the other hand has nearly twice as many questions unanswered as J2. If we break the figure down further by sex, school and type of answer we see that for correct answers the girls of J2 and the boys of J1 have nearly identical mean scores.

If we look at the test items in more detail the pattern becomes clearer. Out of forty separate items only three show any statistically significant differences between the sexes. These are, firstly

Q.2. 'Draw in the lines of symmetry on these shapes.'

Part 1, which required the respondents to draw the lines of symmetry on this shape,

revealed a significant sex difference in favour of boys. A table, broken down by sex and type of answer, reveals:

Table 3: Test items: sex and type of answer

Sex	Right	Wrong	Unanswered
Female	(4)	(31)	(2)
%	10.8	83.8	5.4
Male	(14)	(14)	(1)
%	48.3	48.3	3.4

This is significant at $p < 0.001$ and coincides with the APU Primary Survey Report No.1 (1980a) which gave boys the edge on the symmetry sub-category of the geometry section. However, what is also interesting is that this question was the first of a three-part question on symmetry, the other two parts of which involved drawing the lines of symmetry on the following shapes as well as the one above:

The answers to these last two parts reveal no statistically significant differences between the sexes.

Let us look at the other two items which reveal significant sex differences in favour of girls' correct responses. These were the first part of

Q.6 'These clocks are all ten minutes fast.

Write the correct time under each one.'

None of the other parts revealed any sex differences at all. Answers to

Q.4 '60 per cent of a class can swim.
What percentage are unable to swim?'

revealed a statistically significant greater number of girls than boys getting this wrong, or leaving it unanswered, as did the replies to the question $315 \div 6 = ?$

How can we understand these responses? Firstly, only four out of forty items yielded any sex differences at all. Those there were, were divided equally between boys and girls. In analysing the questions used in *Mathematics and the Ten-year-old Child* (Ward, 1979), Shuard (1981) suggests that girls do better in the questions which teachers think are important, and that in the primary school the teachers tend to rate 'easy' questions higher. By this it can be assumed that computation questions are being referred to. However, our work seems to suggest not that girls necessarily do best in easy or straightforward (computation) questions, but that the way in which differences are assumed between what could be defined as easy or hard questions, or verbal and spatial questions, implies a theoretical basis which is itself open to question (see Corran and Walkerdine, 1981). What seems even more important is that there is little difference in the responses given by the different sexes. In addition, as we shall explore later, it is not the case that teachers in our study valued 'easy' questions in any simple sense.

Equally important is that the between-school differences were actually greater than those between the sexes. Searching for a particular sort of difference can prove to be invidious: anything which points to a similarity then tends to be dismissed as not worth investigation or is written up in such a way as to suggest doubt. In our view that is one of the weaknesses of the interpretations made of the data collected by the Sheffield Polytechnic research team in their report *Mathematics Education and Girls* (1983). Having decided to look at (and for) differences between the sexes they are obliged to try and find them. As they say:

> The reason for the greater emphasis on attitude during the later years of the project was that little difference in boys' and girls' attainment had been found initially. (p.9)

They are clear that:

> . . . the between-sex differences were small compared with the variation within the sex and between the same sex in different schools. (p.31)

This pattern of negligible, often non-existent, sex differences persists throughout the report. Yet the very fact that the research was designed to shed light on what was conceived of as a problem for girls meant that all analyses have to relate to that. They found (p.68) the biggest differences between girls and boys to be their reaction to fractions: '. . . many girls were still at the conceptual level of one-half whereas the boys could understand any fractions'. On the basis of interview data (pp.41 ff.) and of the responses to questions which the pupils worked out separately with the interviewer (pp.59 ff.), they go on to make statements about classrooms:

The girls were more patient in their work . . . This is not altogether a good
characteristic. In a lesson the girls take more care than the boys and so work
fewer problems . . . It is, perhaps, in the reaction to difficulties that, perhaps
(sic) there is a difference between girls and boys. If a girl finds difficulties
in mathematics does she think of the folklore 'girls cannot do maths' and
so gives up, whereas a boy feels it is part of his male image to persevere
. . .? (p.68)

This, we would suggest, leads to the problem of trying to confirm what
is denied by the data. This is not the first time it has been suggested that
actually setting out to find differences often leads the researcher to ignore
data of more importance which may suggest similarities (Maccoby and
Jacklin, 1974, and Gubb, 1983).

It is the relationship between performance and attitude which is at the
heart of the questions raised about girls' performance in mathematics.
For example, the Assessment of Performance Unit's six surveys on
mathematical development, three on the primary sector and three on the
secondary, suggest that whilst at eleven performance differences between
the sexes are slight, and often not statistically significant, it is the case
that girls feel less confident about mathematics than do boys. By the time
of the third published surveys (1982) the APU suggested that, although
the *actual* differences between boys' and girls' mean score on their tests
were slight, they could be seen as foreshadowing larger differences which
appear between the sexes at fifteen (APU, 1982a, pp. 118-9) (but see
previous discussion).

Yet they find it difficult to suggest an explanation as to why this should
be so, other than that girls felt less confident about the subject than did
boys. Earlier in the third survey, however, they had made it clear that
at eleven 'general attitude had little relationship to performance' (ibid.
p.96). Using several different methods we feel we have been able to specify
more clearly the parameters of performance as it is constructed in the
classroom and the social relations within which it is embedded.

Secondary mathematics test*
Again the test devised by ourselves was based on the categories used by
the APU in their five broad categories. This time we also incorporated
aspects of the entry guide to SMILE.

---

* Throughout the discussion on the secondary school data we refer to tutor sets when discussing
the classes in which our samples were located. This has been done to reduce confusion between
class, used to refer to the school class, and social class, as an analytical category. At this
school the tutor set was the basic form of organization.

Using the two guidelines in conjunction with each other a test consisting of twenty-five items, graded according to difficulty and into the areas used by the APU, was produced. Like the SMILE entry guide it was further sub-divided according to levels of difficulty so the first five questions were at level 1; questions 6-11 at level 2; 12-17 at level 3; and 18-23 at level 4, the most difficult. Questions 24 and 25, were miscellaneous applications of number questions and verbal problems. So it was anticipated when the results were analysed that the progressive levels of difficulty of the questions would lead to fewer correct answers. The answers were coded right/wrong/unanswered. Analyses were done by sex and by classes to see what differences different teachers made.

If we first of all analyse the answers by sex there were 96 children in five classes, half the entry intake, 54 girls and 42 boys. Out of twenty-five questions, some of which had more than one part, making a total of thirty-eight items in all, only four items showed any statistically significant sex differences. Three of the items were parts of questions, the other parts of which did not reveal any differences in response.

The first item, the first part of a level 2 question (q.6A), revealed a statistically significant sex difference in favour of girls, 92.5 per cent of whom answered correctly. This consisted of a café price list of twelve different items, such as tea, coffee, cake, chips, etc. Someone was seen to ask for 'Coffee and a cake please.' The question 6A was 'How much will you charge?' This money question involved finding the relevant prices and adding them up. Certainly the majority of the children answered correctly, but significantly more girls than boys. The second part of the question yielded no differences: 'What change will you give from 50p?' There can be no one simple explanation for this sort of response. The question, in both parts, was relatively simple once the price for the items had been extracted from the superfluous information given. It could be that girls felt happier with a familiar, verbal problem involving only the simplest mathematical calculations. The question then arises as to why more boys did not answer the question correctly, given its simplicity?

In all the discussions about differences in response to questions (APU, 1980a, 1980b, 1981a, 1981b, 1982a, 1982b; Shuard, 1981, 1983; Sheffield City Polytechnic, 1983) there has never been any discussion as to why girls do best on 'easy' questions. If these questions are 'easy' (see Shuard's (1981) computation facility levels on questions for 'Mathematics and the 10 year old child'), it would be reasonable to assume that *all* children regardless of sex would do well on them. If, as Shuard says, girls do best on questions the teachers think are important, then why do boys not understand that these questions are 'important', 'easy' and, presumably, worth succeeding at? To divide questions into such categories and then to assume that only one sex is capable of doing well in them, is to beg all sorts of questions about why the other sex — considered better — doesn't do as well. Once distinctions have been made, analytically, between 'easy/hard' then it becomes impossible to explain the mismatch in data. Have boys skipped a stage? Are these questions so easy that they get them wrong? Why use such a dichotomy in the first place? And why therefore is boys' failure so rarely the object of any scrutiny?

The next question showing any sex differences was q.12 which was a level 3 question about angles. Three angles were shown and the question asked 'Which of these angles is a right angle?' 88 per cent of the boys answered correctly as opposed to 62 per cent of the girls. Only 7 per cent of the boys got it wrong as opposed to 30 per cent of the girls. This was a question which could be considered 'spatial', but it also required knwledge of the properties of a right angle.

The first part of q.16, a level 3 question on sets, showed a small difference in favour of girls to the question (based on a Venn diagram) 'Which number is inside all three sets?' However, the second part, 'Which numbers are in just two of the sets?', showed no differences in response between the sexes.

The final item on which any statistically signicant sex difference was observed was q.23. This was a level 4 question (the hardest level) on statistics and it read: 'Five pupils take a test. These are their marks: 23   2   1   17   27. What is the mean mark? (the mean is the ordinary average).'

On this part of the question 79 per cent of girls and 90 per cent of boys attempted it, although 26 per cent and 41 per cent respectively answered it correctly. It was on the second part of the question that the sex difference appeared. This went as follows: 'A sixth pupil takes the same test. The new mean mark is 16. What is the sixth pupil's score?' Only three pupils got it right, all boys, but the difference appears in the response of the girls. 55 per cent left it unanswered as opposed to 27 per cent (just about half) of the boys. Of those who attempted it all the girls (45 per cent) got it wrong to twenty-seven (or 65 per cent) of the boys. so more boys than girls attempted the question even if the answer was incorrect. If we look at the overall results by sex and by the set of the children we see that the overall mean mark was 21.8. The girls' overall mean mark was 21.5 and the boys' 22.1. There are wide discrepancies in the scores between tutor sets — let us list them. Differences from the overall mean

Table 4: Secondary mathematics test: overall results by sex and tutor set

	all five Tutor Sets	Tutor Set 1	Tutor Set 2	Tutor Set 3	Tutor Set 4	Tutor Set 5
Overall mean: boys and girls	21.8	24.5 (+2.7)	20.4 (-1.4)	20.6 (-1.2)	22.2 (+0.4)	20.9 (-0.9)
TOTAL CHILDREN	(96)	(21)	(20)	(19)	(20)	(16)
Girls	21.5	26.2 (+4.7)	20.9 (-0.7)	20.4 (-1.1)	20.9 (-0.6)	19.1 (-2.4)
TOTAL GIRLS	(54)	(11)	(12)	(9)	(12)	(10)
Boys	22.1	22.7 (+0.6)	19.6 (-2.5)	20.8 (-1.3)	24.1 (+2.0)	23.8 (+1.7)
TOTAL BOYS	(42)	(10)	(8)	(10)	(8)	(6)

are in brackets. The strongest group in the first year are the girls in set 1: far ahead of the nearest group to challenge them, the boys of set 4. The weakest groups are the boys of set 2 and the girls of set 5. Tutor-set 2 is the weakest overall and set 1 the strongest. The only real conclusions which can be drawn from this table is that the first-year intake at school S1 was certainly a mixed one. In fact the only conclusions which can be drawn are, as at the primary level, to do with the differences between the set which may relate to differences in teaching.

Looking at the answers to the questions, as they were analysed by set, we see statistically significant differences in the expected directions, with set 1 doing better overall and set 2 worst. Again, as expected, the higher the level of, and consequently the harder, the question, the fewer the right answers.

In answers to level 1 questions set 2 showed the most difference, leaving q.3B (whole number computation: ? - 16 = 34) unanswered and splitting 50/50 right and wrong answers on q.5, a fractions question: 'What fraction of the shape is shaded?' On the next two items — a generalized level 2 arithmetic question (? × 8 = 96) and a symmetry level 3 question (two lines shown with a diagonal labelled 'mirror': 'Draw the reflection of the lines in the mirror') — show set 2 getting significantly fewer right answers than the others.

As in the primary data it would appear that what marks differences between pupils tends to be their reaction to questions which look unfamiliar, as to whether they are tackled regardless of the outcome or they are left unattempted. This would seem to reflect a difference in confidence, a willingness to take risks, which seems to have more to do with

what is normally expected of the pupils in their classes than with sex. It
could be related to social class, but our evidence is too scanty to show
much of a relationship.

Fourth year secondary mathematics test
The fourth years were tested during the terms in which we did our fieldwork
as a preparatory measure to deciding which examinations (O level or CSE)
would be taken by which pupils. We decided to analyse the results of the
school mathematics examination for several reasons: (i) the school were
not keen to give the children another test so soon after this one; (ii) the
test was wide-ranging and covered quite similar areas to the APU secondary surveys 1 and 2; (iii) the pilot nature of the fourth year work and
the smallness of the sample in the fourth year meant that we had little
or no knowledge of other fourth-year classes other than the one in which
we carried out the fieldwork. These test results provided us with a wider
data base.

The test given was divided into two sections, A and B. Section A was
made up of twenty questions covering, in no particular order, all the areas
of mathematics. Section B was divided into six sections, all dealing with
specific areas of mathematics such as Sets, Cubes, Triangle Transforma-
tion, Money Matrices, Centigrade and Fahrenheit, Journeys. As it was
marked for the school's purposes the first section was worth 40 marks
and the second 60. In section B the four best marks out of the sections
were taken. The mark out of 100 was then scaled down to give a mark
out of 40. A total mark was then achieved by combining (i) the exam mark
(ii) mark for course work out of 40 and a mark out of 20 for investigations. Since it was preparatory for public examinations and so designed
to be held in similar circumstances, this test was timed. However, for our
purposes we re-marked the scripts for 5 out of the 10 fourth-year forms
giving each answer a score (as for the first years) of right, wrong or
unanswered. There were 62 boys and 35 girls whose scripts were marked
by us. This does not include all the children in this half of the fourth year
since 15 boys and 6 girls were persistent non-attenders and had not taken
the examination at all. Unlike the first-year and fourth-year junior samples,
in the fourth year of the secondary school boys outnumber girls by almost
2:1 (76:41), whereas, in both other age groups we sampled, the figures
were reversed in favour of girls.

When the printouts were analysed for type of response by sex there were
no statistically significant differences on the first part of the test. On

section B, the section on sets, the last three questions, 14, 15, 16 showed proportionately more girls than boys answering the questions, regardless of whether the answers were right or wrong. The answers tended to yield no statistically significant differences by sex, rather by type of response. Certainly in the second section the last questions of each part yielded more unanswered questions. Of the temperature questions only one part yielded any difference and that was to convert O degrees centigrade into degrees fahrenheit. There was a statistically significant difference between sexes in that more girls than boys left this question unanswered. There were five questions in this section on conversion of temperature but only this question showed any significant differences.

Looking at the analysis of questions by tutor set there were far more statistically significant differences. On the cubes section of section B set 1 mostly got the entire section wrong, suggesting a lack of familiarity with the area of 3-D diagrams, mapping diagrams, volume, area, etc. In fact the most differences are those between set 2 and set 1 in terms of right and wrong answers, with set 1 lagging behind.

We were also provided with details of the marks given to the pupils in the other parts of the total work marked. The final mark in this examination was made up not only of the examination result but also of the results of work done in class (course work and investigation work, open-ended problem solving). The tables which follow show the respective marks by sex and set of the fourth year. Firstly, the average marks for all the areas examined (average marks given).

Table 5: fourth-year secondary mathematics test: marks by sex and tutor set

(total figs in brackets)

	Overall Average	Tutor Set 1	Tutor Set 2	Tutor Set 3	Tutor Set 4	Tutor Set 5
Boys and girls	57.9	48.9	63.9	59.9	62.3	56.5
Total 97		(23)	(19)	(15)	(23)	(17)
Girls (35)	62.5	56.9	58.4	74.5	68.1	59.8
		(10)	(7)	(4)	(9)	(5)
Boys (62)	55.3	42.8	66.1	54.5	58.6	55.1
		(13)	(12)	(11)	(14)	(12)

These marks include the marks for course work (out of 40), investigations (out of 20) and examination marks (out of 40). As mentioned earlier there were 35 girls and 62 boys. These marks show that overall girls did substantially better than boys and they scored higher marks in all sets except set 2 which contained a group of boys who scored the highest overall mark. Tutor set 1, as anticipated from the results of the analyses of the examination results, had the lowest overall mean score.

The results for the investigations separately are set out in Table 6.

Table 6: fourth-year secondary mathematics test: investigations marks

	Overall average	Tutor set 1	Tutor set 2	Tutor set 3	Tutor set 4	Tutor set 5
Boys and girls (94)	12.4	12.2 (22)	14.2 (19)	11.7 (15)	11.9 (23)	12.1 (15)
Girls (34)	14.4	15.2 (10)	14.4 (7)	14 (4)	13.6 (9)	14.8 (4)
Boys (60)	11.3	9.8 (12)	14 (12)	10.9 (11)	10.9 (14)	11.1 (11)

As can be seen one girl and two boys were absent for this work. Again the girls in all classes did better than the boys and the four girls in set 3 have high scores. Table 7 for course work shows this pattern.

Table 7: fourth-year secondary mathematics test: course work marks

	Overall average	Tutor set 1	Tutor set 2	Tutor set 3	Tutor set 4	Tutor set 5
Girls and boys (97)	24.3	22.5	25.6	24.3	26.3	22.5
Girls (35)	26.3	25	24.3	30.3	29	23.6
Boys (62)	23.2	20.6	26.3	22.1	24.6	22.1

Again, the boys from set 2 score the highest score for boys but the girls of set 3 score an exceptionally high score out of 40. Apart from set 2 the girls consistently outrank the boys. The final set of marks are for the examination results which we have looked at briefly. Broken down by set the average scores are set out in Table 8.

Table 8: fourth-year secondary mathematics test: examination marks

	Overall average	Tutor set 1	Tutor set 2	Tutor set 3	Tutor set 4	Tutor set 5
Girls and Boys (97)	21.9	16.6	22.7	23.9	24.5	23.3
Girls (35)	23.5	21	19.7	30.3	25.6	24.4
Boys (62)	21.1	13.2	24.4	21.5	23.8	22.8

Table 9: fourth-year secondary mathematics test: scores on all results by class and sex

(The total scores refer to the results of all the items added together)

		Investigations	Coursework	Examinations	Total
Highest score GIRLS	Tutor Groups: 1	3	3	3	
BOYS	Tutor Groups: 2	2	2	2	
Lowest score GIRLS	Tutor Groups: 4	5	2	1	
BOYS	Tutor Groups: 1	1	1	1	

Table 9 shows which sets did best and worst on all the dimensions examined. The boys' scores are clear — those in set 2 came top in all areas and the boys in set 1 came bottom in all areas. The girls in set 3 did best in all areas except investigations — which could be considered the most open-ended and least structured area. The girls from set 1, whose overall total marks were the lowest of all the groups of girls, came top in that area. Otherwise, the lowest marks in the other areas were shared by other sets — each one presumably showing one particular weakness.

The reason for presenting the data in this way is to make the point that *alone* the test data provides no complete explanation of the scores. It would be difficult to make statements about how these scores are produced if we had no data on classroom practices and the constitution of teachers and pupils within them. If we are to consider intervention it is important to know what goes on in classrooms (the end result of which are the test scores) as well as the effects for policy.

In December 1982 the fourth year took a mock examination to decide for which public examination they would be entered. Out of the entire year's intake of 229 pupils 31 children were entered for O level. Of the rest all but ten were entered for Mode 3 CSE. The remaining ten were entered for Mode 1 CSE. The difference between CSE Mode 1 and Mode 3 is that the Mode 1 is set and marked by the examining board. Mode 3 is set and marked internally by the school and moderated by the board. In this particular case the examination was based on the SMILE scheme and co-ordinated centrally by the SMILE Centre. Mode 3 was the examination most frequently chosen since it depended on course work as well as examination performance. Mode 1 was suggested only to those children who, for whatever reasons, had not done much course work.

In the school's opinion, the CSE Mode 3, since it incorporated work done in class with that done in examination, gave a fairer assessment of competence than did Mode 1 or O level. However, despite these very valid reasons O level is still the most prestigious examination and more acceptable to future employers. Only Grade 1 CSE is acceptable as an equivalent qualification.

The average mark needed at this school for entry into the O level set was 82.5 (three girls and seven boys were entered despite having lower than the average marks). Of the twenty-eight (out of thirty-one) children whose sex was known (by us) who were entered for the O level twenty were boys and eight girls. Despite the fact that, in the mock examinations which we have analysed in this chapter, girls performed overall better than boys, in four out of five tutor sets, hardly any girls at all were entered for O level. It is certainly the case that the number of girls being entered for O level is rising and that girls, once entered, tend to pass. Our interpretation of the data suggests, conversely, that *despite* their success in the fourth year relative to boys, girls are not being entered for O level in anything like the same proportions as boys.

While it is quite common to treat test results as though they presented a picture of incontrovertible fact about performance and, by implication, ability, these results suggest that something is happening in the school in which superior performance on the part of girls is being downplayed in producing practices which discriminate against the girls' entry of O level. As we have suggested earlier in the chapter, it is not uncommon to utilize very slippery criteria when interpreting test data.

We shall go on, in the next chapter, to demonstrate that the arguments used by teachers for putting girls in for CSE rather than O level are based upon characteristics which are displayed in the classroom which the teachers tend to interpret as 'lack of confidence', This has the consequence of making such teachers' feel that they should protect rather than push the girls in question. However, we should hardly be surprised to find that, as a consequence, girls' CSE results are not bunched at the top end of the attainment range. Our emphasis then is to suggest that, although test results are often understood as 'hard' and classroom evidence as 'soft' data, if we ignore the classroom practices which help to consitute what good and poor performance means we are in no position to have any sense of the complex causality involved in the production of test and examination results.

Summary

Let us summarize the conclusions which may be drawn from our analysis of the test data from the fourth-year primary, first-year secondary and fourth-year secondary samples.

1. There are very few statistically significant sex differences in the primary and first-year secondary tests. There are, however, large differences between schools in the primary test and between tutor sets in the first-year secondary test.

2. In the fourth-year secondary test girls consistently outperform the boys. This is the test used by the school to plan whether children are to follow courses for O level or CSE.

3. There is no support in these data for the commonly-held view that girls simply perform better on computation or low-level items. In respect to such claims, the failure of boys on low-level items is never questioned. If the items are simple, why are boys not able to do them? As is commonly found in gender research it is girls' results which are taken to be a problem.

4. Despite the high performance of girls in the fourth-year secondary tests it turned out subsequently to be considerably more difficult for girls to be entered for O level mathematics than boys. Boys appear to be entered *despite* certain adverse characteristics (such as 'haphazardness') whereas girls are less frequently entered despite high scores and favourable evaluations such as 'conscientious' and 'co-operative.' (Although a Grade 1 pass in Mode 3 is taken as an equivalent for O level in practice it is not equivalent at all.)

On the basis of these conclusions certain inferences may be drawn. Firstly, although it is standard in the literature to make extrapolations about classroom performance and ability from test data, we suggest that inferences are often made much too starkly. The differences in performance between schools and tutor sets leads to the supposition that the social relations, teaching and learning in *specific* classrooms, are of considerable importance in accounting for test attainment. Moreover, despite the expectation that girls' performance should have declined by the fourth year, in fact girls outperform boys and are still less frequently entered for the higher status examinations. This suggests that rather than the 16-plus examination differences between girls and boys being able to be accounted for by a simple decline in performance, there is a complex relation between teacher evaluation, classroom performance and examination entry.

The data reported in this chapter, then, do not support existing assumptions about the mathematical failure of girls in any simple sense. They point to the necessity of examining in detail the actual practices of the classrooms in which performance is produced and the evaluations and judgements made by teachers. It is these issues which we shall now address.

Chapter Four
Teacher Interviews

This chapter will examine the data collected from teacher interviews. Given our emphasis in earlier chapters on the importance of classroom practices and teacher evaluation, our analysis here will focus upon the issue of practice and judgement. What terms and categories do teachers use to describe themselves, their practices and their pupils? From where are such understandings derived? How do they affect and effect what happens in their classrooms and therefore the performance of their pupils? How far and in what way are evaluations and practices similar and different, from teacher to teacher, from school to school, from primary to secondary school?

In the introductory chapter we referred critically to the concept of the 'hidden curriculum' and its importance in raising the issue of the classroom production of gender. The implication is that processes, usually unexamined, are at work beneath the apparent curriculum which do much to shape the kind of behaviour and responses produced by girls and boys. Areas which have been studied to show the working of the 'hidden curriculum' are, to name but a few, practices of segregation of girls and boys, illustrative material in books, and patterns of interaction. We feel that such approaches are limited precisely in their splitting off of the hidden from the *overt* content and processes of the curriculum and schooling. They leave unexplained the effects of the overt practices, the conceptions of learning, classifications, organization and content of schooling, which should not be separated from consideration of the more informal processes.

Let us give an example of what we have in mind. The hidden curriculum approach would examine, for example, 'sexist bias' in the content of the mathematics problems on which the children in each classroom worked. From such information would be extrapolated processes taken to produce norms of feminine and masculine behaviour, roles and sterotypes. Related behaviour demanded by teachers might well be similarly stereotyped. However, while it is important to examine such processes (an analysis of textbooks has been carried out under the auspices of the Girls and Mathematics Unit: Northam, 1982), focusing on content alone leaves untouched other issues about the mathematics curriculum. In the previous chapter we pointed out the commonly held assumptions about girls' superior performance on low-level computational questions and boys on higher-level problem solving. We alluded to the debate about girls being capable only of the activities belonging to a conceptually lower order. While we suggested that the evidence did not in any simple sense support such conclusions, it is important to examine why the distinction between rule following and proper conceptualization is one which (a) has considerable force and (b) is applied to girls.

It can be argued that central to some psychological assumptions about learning and cognitive development, and to modern theories and practices of mathematics teaching and learning, are certain important distinctions. Examples of these distinctions are 'rule following' and 'proper conceptualization'. They have significant and important effects for defining what counts as proper attainment. If girls are judged successful *but* their success can be said to be founded on rote learning and not proper conceptualization, then certain adverse consequences follow from this mismatch of attainment and explanation. Clearly it is not just a matter of hidden processes reinforcing roles and stereotypes. Rather an examination of theories, practices and teacher judgements is needed in order to understand the complex interplay of conditions by which girls and boys' performance is produced, monitored, evaluated and regulated. We have written about the interplay between developmental psychology and mathematics teaching elsewhere (Walden and Walkerdine, 1983; Walkerdine, 1983, 1984). Here we can only make brief reference to such arguments, though they impinge directly on our analyses of the data.

The interviews were carried out with all the teachers as well as the children. Our teachers filled in questionnaires prior to their interviews about their backgrounds and thoughts on teaching mathematics to give us some common ground for discussion on a general level, rather than about any specific set children.

We asked all the teachers (three in the primary school and ten in the secondary school) to rank the children in their classes in terms of whether they considered them to be good and poor at mathematics. This provoked some hostility and not a little anxiety on the part of all the teachers and at all levels. They did not wish to commit themselves by an admission that they did in fact classify children. The work which has been published on teacher expectation had obviously influenced them in this. However, despite expressions of concern, which we noted, no one actually refused. The classifications given form an important part of our data when related both to classroom practice and the children's views of themselves as learners of mathematics.

Primary teachers
The interviews with the teachers whose classes we observed revealed that they saw mathematics in the primary school in much the same way: conceptual understanding was the bedrock of primary mathematics: the main aim should be to provide tools, lay foundations, for later life.

Both female teachers (one from each school) saw differences between the boys and girls in their classes. Ms C. from J2, talking about one of the girls she felt was poor at mathematics, said: 'She is very dicey on maths in that she just doesn't have any confidence at all.'

RW: Do you think confidence is important in tackling maths?

Ms C: Yes, very important. I think you've got to be confident about your . . . about the situation you're in and how to handle it . . . in terms of problem solving . . . you can't take off at all, start thinking necessary to solve the situation when you don't have confidence. (Lack of confidence) seems to stop people starting their thinking pattern in maths. You might be able to start it in writing things down and cover that up so they don't get into the subject properly, but not in mathematical situations.

What then were her views on differences in ability between boys and girls?

Ms C: No, I wouldn't have said outstanding differences, but what I have found interesting with this group of children is that at the end of the third year the boys started to think in more abstract form.

RW: What do you mean, they began to think in more abstract form? How did you notice it?

Ms C: When we did the calculator work they were able to think in terms of what you were doing with the numbers rather than in terms

of what added on to what. They were more interested in the processes
. . . than worrying about whether they got it right, and to be able to
be free, to free themselves from that . . . and it was natural.

Ms A. from J1 had similar feelings about the girls in her class. She
thought girls worried more: they were academic, capable and hard working, but tended to be more anxious. In her view 'they perpetuate the
stereotypes by choosing to do tasks, to help'. Boys on the other hand were
'more creative, divergent and tangential . . . see more opportunities to
explore'.

When asked to think about the bright boys and girls in her class and
then the slow ones, and to think about differences and similarities amongst
them, Ms A. replied that the highest-scoring boy and girl (on our test) were

sort of convergent thinkers. Very much directed. Get on well in formal
directed atmospheres. She is terribly diligent, but she (the next highest
scorer), she's basically better at maths because she thinks and she's
interested and she finds it fascinating and she's interested in the odd
bits of it. She's untidy and slap-dash but she does things. But she's
(first girl) very nervous about maths . . . she always finds security in
the formal bits, computation and patterns and her work is always
beautifully meticulous and typical of her personality.

Both these teachers (Ms C. and Ms A.) tended to see boys as more
interesting, annoying certainly but with that extra spark that the hardworking anxious girls lack.

A common theme which can be picked up from the responses of these
two teachers is the counterposing of, on the one hand the active, enquiring, rule-breaking child with the well-behaved, passive, rule-following
child. The ways in which teachers phrased their responses almost always
meant that active children were boys and passive ones, girls. This, in fact,
is the way in which the argument about girls' performance has been couch-
ed — and one of which we have been critical (see Shuard, 1981). It is closely
linked to arguments about progressivism, learning by doing and discovery,
etc.

The secondary school
The school (S1) in which we conducted the next phase of our work has
a long commitment to comprehensive education. It was amongst the first

in London and the school ethos is heavily in favour of mixed-ability work and the ideals of a balanced and comprehensive education for all. It has a very widely mixed catchment area in terms of the diversity and range of both the class and ethnic backgrounds of the pupils. The school tries to ensure that the ten tutor sets of the first year are composed as far as possible to allow for an adequate and representative selection of different sexes, social class and racial backgrounds and ability groupings.

In this school the mathematics department was committed to the teaching of its subject in mixed-ability groups up to the fifth year. Then, however, it selected O level and CSE groups. There was a move afoot to continue mixed-ability teaching up to the examinations with different provision, within the same class, for pupils taking different examinations.

Inherent in the idea of mixed-ability teaching is the view of pupils as individuals with different rates of development and different needs. Rather than streaming into different groups according to ability — as defined by some sort of test — there is a wish to prevent this labelling — seen as an inevitable corollary of the testing process — and to provide more flexibility for both teachers and pupils (Kelly, 1978).

However, as we have seen in the primary school (Walden and Walkerdine, 1982; Eynard and Walkerdine, 1981) and shall see further in Chapter six, to think of children as individuals with individual needs demands a re-thinking of teaching methods. It becomes impossible to teach a whole class the same topic in the same way because of the diversity of ability. Our secondary school was involved, almost from the beginning, with the development of new methods evolved specifically to deal with mixed groups.

Throughout the school mathematics was taught using SMILE (Secondary Mathematics Individualized Learning Experiment). This acronym is indicative of the orientation of those involved in its development. who wished to make mathematics enjoyable. SMILE's inception came in 1972 after a meeting with the originator of the Kent Mathematics Project — the first mixed-ability scheme for teaching mathematics in comprehensive schools — of several heads of mathematics departments in London schools.

The scheme is devised on a matrix comprising all the areas of mathematics to be covered at varying levels. Teachers can choose to work horizontally covering different topics, or vertically, pursuing one topic in depth. All areas are covered by a graded series of work cards whose numbers are noted on an individual matrix comprising ten tasks which each child in a class is given.

Record-keeping is an important part of the teacher's task: monitoring progress and development of expertise. It is also an important part of the child's role in that each child is expected to mark and self-correct work. Only when this is done for each task can the child be given new work, and 'signed off'.

Comprehensive schooling seems to have been wholeheartedly accepted as the best social form of organizing secondary education by those at the 1972 meeting. However, as at other times, any fundamental social re-organizations were met in certain quarters with criticisms about falling 'standards' particularly in those areas seen as the 'basics'. The teaching of mathematics is notorious for coming under fire at these crucial times. Consequently, the literature on SMILE which began to appear from 1973 onwards was concerned to give a teacher's eye view of the scheme and its rationale. The main strands are that:

1. Children work at their own pace and at work suitable for them,

2. The teacher guides them through it,

3. The scheme is flexible and in a constant state of development.

The old 'chalk and talk' methods were felt to create passive pupils waiting to be fed: true to its, unacknowledged, theoretical background in cognitive developmental theory, SMILE's aim was to motivate children to work because they were enjoying it and could take responsibility for it. Ronnie Goldstein, the originator of this scheme, wrote in December 1973:

> . . and when they (the pupils) have learnt to use it properly they enjoy the total responsibility they are given for organizing their own activities. (Goldstein, 1973)

One of SMILE's strengths was felt to be its ability to 'free' the teacher from mundane organization to being able to teach. As Goldstein put it:

> . . . with the teacher relieved of all routine classroom organization he is freed to attend to the rather more important matter of education — tutoring individuals and also temporarily-formed groups. (ibid.)

Signing off (when the child has completed all ten tasks on the matrix) means the teacher can monitor work, but errors rarely happen:

. . . because the teacher is circulating during the lesson and tends to spot the type of mistake that would lead to drastic errors.

As Langdon saw it, the role of the class teacher was '. . . to direct the child to the point of discovery'.

The warden of the Ladbroke Mathematics Centre, where the scheme was devised, developed and still has its headquarters, wrote in October 1975 that the 'teacher's role was that of adviser rather than a dictator'. Rachel Gibbons saw one of the scheme's strengths as being that it had enabled the teachers involved

. . . to grow in stature as teachers learned to take criticism and offer it constructively, become more self-critical in their work and of the quality of the material they put in front of their pupils. (Gibbons, 1975)

The introduction to the SMILE scheme states:

The project features individualized learning, mostly with mixed-ability classes of thirty children, with each child: (a) working on one of 1,400 tasks; (b) working at her own level of ability; (c) working at her own speed; (d) choosing, with an assignment, her own order of working; (e) marking her own daily work; (f) working in a group when appropriate; (g) writing an appropriate test after each assignment; (h) responsible for materials/equipment needed.

The teacher's role varies tremendously according to personality but the common features are to: (a) provide a working environment with learning materials and equipment; (b) develop the child's responsibility for her own work; (c) check how pupils have done on individual tasks; (d) teach to a group or the class if appropriate.

Additional work for the teacher outside the lesson time involves: (e) marking a pupil's tests on completion of an assignment; (f) choosing a new assignment (usually ten tasks on a variety of mathematical topics recorded on a 'matrix' on the basis of the test and classroom contact).

It is seldom that one would see a SMILE class: (a) having a formal, teacher taught lesson; (b) working on the same topic; (c) working at the same level of difficulty; (d) with no movement in the room.

. . . efforts are made to provide tasks that are suited to each pupil's ability and experience. The emphasis is on the pupils' learning from activities that they carry on independently, using the teacher as one resource amongst many, and not on formal teaching and teacher-directed class activities (ibid.).

So what we can establish is that teachers operate within particular frames of reference which derive from how they conceive of learning in the classroom. In turn, this can be related to the rationale of the methods used.

Secondary teachers
Primary schools are the site of concern as the child's first introduction
to education, and there is concern to teach the rudiments of all areas of
knowledge which can be used as a foundation on which later education
can build. But secondary school is a different matter. This is where 'real'
learning is often taken to begin. The day is differentiated into subject areas,
with different teachers, usually in different places which are specially
designed for the purpose. The curriculum becomes strongly circumscribed — syllabuses abound and need to be worked through. The training
of the teachers is often different. Whereas primary teachers tended (this
is changing as the structure of teacher training changes) to come from
CertEd or BEd courses, with an emphasis on child developmental
psychology, secondary teachers were specialists in their own field in which
they had a degree and a postgraduate qualification in education. PGCE
courses are more intensively geared to teaching practice with the possibility
of less educational theory than is provided on full-time, four-year BEd
training courses, but every teacher with a teaching qualification has some
background in educational theory. Mathematics teachers, however, often
have no teaching experience or qualifications, but are accepted without
these because expertise in their subject is in short supply. Thus for these
teachers any educational theoretical orientation may have been acquired
on an ad hoc basis.
   The Cockcroft Report (DES, 1982a) categorized the mathematics staffs
of 500 maintained secondary schools surveyed into four levels of qualification, based solely on academic criteria, with the caveat that paper qualifications do not automatically imply a good classroom teacher.
   The four levels of qualification were:

1. Good — basically anyone with a first degree in mathematics or a
    related subject plus a PGCE or BEd with mathematics as a main
    subject.

2. Acceptable — mathematics graduates with no teaching qualifications
    or those with CertEd in which mathematics was a main subject.

3. Weak — graduates or those with a CertEd with mathematics as a sub-
    sidiary or those trained for primary schools or those with a degree
    in mathematics related subjects.

4. Nil — teachers with no mathematics specialisms or without degree status in any mathematically related subject. (ibid.)

If we look at the qualifications of the teachers we interviewed we can make some suggestions as to where their perspectives on teaching and learning mathematics had been derived. In fact of the nine teachers seen in the department three had 'good', four had 'acceptable', and two had 'weak' levels of qualification. Six out of the nine had pursued some form of teacher training, usually at postgraduate level.

We asked all the teachers about their opinions of the SMILE scheme, as well as about the differences they could observe between the good and poor children in their classes and the boys and girls. Taking the groups of teachers separately we see that the group of 'good' teachers, all of whom had mathematics degrees and PGCE qualifications, all felt quite positive about SMILE, although one had serious reservations which will be explored later. All three mentioned the value of SMILE as being that it began from the individual's own standpoint and enabled a programme to be developed which would suit each child.

. . . the emphasis is on where the pupil is at, how the pupil thinks, where the pupil and teacher think he/she should be developing from their own personal starting point. It also gives more credence to an individual way of thinking . . . Most important, strength is freedom: to take a fresh and individual view of the subject . . . to experiment etc. . . . Encourages a 'learning through doing' approach to maths teaching.

These quotations, taken from responses to questions about SMILE, demonstrate clearly that underlying this point of view is a conception of child-centred education as exemplified in the primary school: learning through doing, with each child having an individual starting point.

These three teachers found the only problems with SMILE — as an individualized learning scheme — to be administrative, i.e. equipment was missing, cards were written in difficult language not easily understood, lazy teachers could use it to introduce mixed-ability teaching without con-sidering what that meant. Throughout the replies ran a wholehearted com-mitment to mixed-ability teaching in mathematics. The commitment to mixed ability in the secondary school is a development of the child-centred approaches of the primary school and is based on the rationale that it is possible to cater for all needs in one class by providing graded and

structured materials to suit all abilities. (This is why SMILE is divided
into various levels both of topics and of difficulty). Social reasons are
also important.

As one of the group of teachers put it, mixed-ability teaching is seen as,

> . . . providing the possibility of different abilities interacting to
> mutual advantage, the bright children learning more through explain-
> ing to others, children not being irrevocably stereotyped.

This teacher had reservations about the scheme and his criticisms help
to clarify the foundation on which the scheme was built. We shall quote
from him in full and then discuss his statement:

> The illusion is that children learn from reading the work card set,
> testing their understanding by answering the questions . . . All too
> often children ask for the teacher's help, not having read the card
> or . . . not able to read the card with understanding . . . children
> want to know 'how to do the card' not learn what it means and how
> it fits in with previous work. The cards are seen as individual tasks,
> the overall rationale or topic is rarely perceived. . . . Children's reten-
> tion is diminished through a failure to teach them to communicate
> what they are doing . . . this verbal experience is essential but not
> part of the scheme.

The underlying rationale of the scheme is that children learn better if
allowed to 'discover' for themselves. Like the primary teacher therefore
the mathematics teacher becomes a facilitator, a provider of context, and
not a leader or instructor. However, as in the primary school (Corran and
Walkerdine, 1981), what becomes important for the children is to complete as many cards/matrices as possible and learning becomes secondary
to accumulating matrices. Also by fragmenting topics there is little continuity for the children who may not recognize similar areas of work in
different contexts. As in the primary school, discussion although not
discouraged is not actively encouraged, so children have little opportunity
to explain areas and make sense of the work.

To sum up, then, the teachers with the 'best' qualifications rely on a
psychological model of the child as learner which presupposes that
knowledge is internalized activity. The model draws for its overall rationale
on the encouragement of mixed-ability teaching and 'progressive' ideas
of education. It aims to privilege the child as learner with different sets

of experiences, abilities and aptitudes all of which must be capitalized on. How do these teachers classify the good and poor pupils in their classes?

All of them characterized the good children as confident: . . . '(They're) not afraid of getting something wrong . . . They're prepared to try an answer, because if it's wrong then they just try the next one, or another approach.' 'Confidence', 'perseverance', 'solid background' are all adjectives applied to the good children — boys and girls. '(Good children) have the ability to get on by themselves and also initiate new ideas.' In addition they have solid foundations of expertise, and are interested in the mathematics for its own sake. The teachers had a liking for children whom they perceived to be interested in the subject for its own sake. For these three teachers their most outstanding pupils were all girls. These are the terms they used to describe them:

(She) would be largely talking about maths . . . the main part of it would be to learn something about the maths.

(She's) intelligent and intellectually confident . . . lots of good ideas and initiative.

(She) can do anything given to her, a really outstanding mathematician.

The poor children were characterized by a lack of confidence and an inability on the part of the teachers to distinguish them from each other clearly. They were seen to need a lot of support. One teacher summed up the feeling about the poor children as:

They're actually kids who don't think of themselves that their job in life is to grow and learn and that that's possible for every human being. They think there's another game to play in which they're at the bottom of the ladder and therefore they don't engage in the game.

Behaviour became a criterion for judging achievements and those children who were disruptive were felt to achieve least and be most in need of help. One boy's achievement was seen as 'a function of how much help I give'.

The child-centred approach adopted by these teachers leads them to see failure or success as a product of individual biography/psychology. There is little recognition of the social setting and pressures other than those which are very noticeable, for instance disruptive behaviour or poor language skills.

The four teachers in the 'acceptable' category also mentioned the usefulness of SMILE as an integrated programme of individualized learning which was ideal for teaching a mixed-ability class. The role of the teacher was mentioned: 'SMILE allows the teacher to become an adviser rather than being the sole source of information and knowledge.' Freedom and autonomy for pupils was another plus and included the freedom to choose topics, work rate and peer groups. These teachers then reiterate the feelings of their colleagues about the strengths of the system and seem to echo the child-centred primary teachers in their psychological and individualized view of learning.

Criticisms of SMILE from this group had little to do with the learning process. Again it focused on the administration of the scheme. The main complaint was that if used by a teacher who did not understand it or was not committed to it (presumably to the model of learning it embodied rather than the cards per se), it could, used without a break, become boring. Again, the complicated language was criticized and the fact that it was not possible to do class or large group work, since not every child would have done the necessary ground work.

For these teachers the good children were distinguished by their ability to understand and pick up explanations or concepts quickly. They were well organized, thorough and systematic. One teacher felt that the good children were distinguished by having very solid mathematics backgrounds both at their junior schools and at home. He considered the home background important: '. . . that helps and will stay with them for a long time. Gives them a tremendous start.' Poor family background was an excuse for poor achievement. One teacher in this group contrasted those children with flair to those who worked hard, particularly girls. On the one hand there were chldren who

seem to have a natural flair for it. They seem to be able to argue the concepts out with themselves and come up with some interesting ideas.

On the other the girl who

. . . works damned hard . . . it's obviously through damned hard work that she's discovering.

All the teachers seemed to counterpose the interested children with flair to those who were workhorses, or didn't have the interest or ability. The poor children were considered disorganized and slow to understand: '(of) low ability at mathematics anyway'.

Just as this teacher compared girls in terms of flair and hard work, two of the four teachers discussed them in terms of their presentation. One of the teachers, who considered organization extremely important and who differentiated between the good and poor children in his class on their ability to organize and write down their work in a systematic fashion, took his analysis one step further and claimed that poor children tended to over-accentuate their presentation: 'to compensate for lack of mathematical understanding'. As he put it about one of the poor girls who he claimed overdid the presentation: 'all that kind of detailed work, of course, hides the main mathematical concepts'. A good boy was considered still to be good even though: 'he won't bother about presentation but I can see that he understands the mathematical ideas and he proceeds very well on that basis'. And that from a teacher who claimed that the ability to write work down systematically was very important in being able to make sense of it.

The other teacher concurs about the overemphasis on presentation but also notices what was mentioned earlier, which is that some of the girls considered poor just '. . . tend to rush through . . . they just want to get the matrix finished and carry on with another one . . . they give up very quickly . . . and they can't be bothered to sit down and work it out carefully.'

Lastly were those teachers in the 'weak' category who both had degrees in subjects related to mathematics but no specific teaching qualifications. They mentioned the importance and value of SMILE for mixed-ability teaching. Its flexibility 'and its ability to usefully employ most children at a level of work suitable for them' was mentioned along with the fact that the weaker children were not exposed nor stigmatized for not 'keeping up'. As for weaknesses, the organizational problems again cropped up but with a different emphasis. Whereas one of the other teachers had welcomed the teacher's role being reduced to that of an adviser, a teacher in this category considered the reduction of the teacher's role to that of an organizer to be a problem, especially since that implied that SMILE was invested with total authority. The scheme needed more than to be just handed out to children. It needed to be thought about and topics which were not well structured in it taught differently. Thus there are teachers who do not necessarily accept uncritically the rationale of SMILE. They feel it has weaknesses, even if these weaknesses are not considered irremediable. This could be because the teachers have not had access to the 'educationalist' discourse about learning theories which informs the

practice of those teachers with some form of teaching qualification.
Their categorizations of the good and poor children in their sets were
different. One thought that the good children in his set had all come from
good backgrounds from which they had gained their knowledge about
mathematical concepts and were able to pick up and capitalize on these
quickly. The other teacher saw the good children as confident and enthusiastic if not always wholly accurate. He felt disappointed by the
children from middle-class backgrounds as he did not feel that their initial
promise was actually borne out in practice. One teacher described the poor
children in terms of their mathematics ability, i.e., they had poor presentation and writing, and they failed to understand the terminology or to
relate their skills to similar problems in unfamiliar contexts. The other
described his poor pupils in terms of their social backgrounds: 'socially
isolated', 'social problems in relating to people so severe that they distort
(her) potential'.

It seems that teachers with least access to educational theories use
common-sense theories and administrative categories to describe their
pupils.

Fourth-year teacher
The categories used by Mr N to describe his fourth-year mathematics set
were similar to those he used to describe his first-year group, and similar
to those used by other teachers. Good girls, for instance, were seen as
conscientious and confident, able to grasp new ideas quickly. They were
seen to have good backgrounds in mathematics which had stood them in
good stead. A word which was only ever applied to good girls in the context of their mathematics lesson was 'ambitious'. It was used on two
separate occasions to describe good girls and seemed to be hinting at their
motivations for being good, although there was no real clarification from
the teacher. His description of one of them gives the essence of what he
felt about the good girls in his class:

   very conscientious, very well motivated, ambitious, wants to do well,
   not terribly interested in the subject for its own sake. Just tremendous perseverance. Lots of support from . . . family . . . so is pretty
   successful.

Good boys were seen as good at 'visual things'. Again, they benefited from a good mathematics background and were seen as having 'flair' and 'natural ability' — seemingly unlike the girls who are 'conscientious' and 'well motivated' but not necessarily endowed with 'flair'. Good boys were also seen as being interested in mathematics for its intrinsic worth rather than for any instrumental gain, such as an O level or CSE grade 1 pass. The teacher here contrasted a good boy in his class, whom he described as,

[A]stereotypical boy, you know, interested in science and so maths and aeroplanes and stuff like that. So would see maths as important and intrinsically interesting.

with poor girls who were unsure and unwilling to take risks. In fact, the stereotypical girl was seen to be

. . . very unsure of herself, unwilling to volunteer things that might be wrong, wants to take it away and look at it carefully and spend some time going over it again before actually committing herself . . . she'll apparently have no confidence in her ability.

Poor girls lacked confidence and needed lots of help and support to grasp ideas. Poor boys, on the other hand, tended to bluff and cover up their faults. They were considered lazy although some were excused because of the poor self image they derived from having a poor family background.

Summary
In the characteristics mentioned by these teachers as differentiating the good and poor children in their classes there is a distinction which runs throughout our educational system. There are children, inevitably considered good, who have what is variously characterized as 'flair' or 'natural talent' and other children who 'work hard' or 'need attention'. The dichotomy is often not expressed quite so starkly but it is a distinction exemplified in most of the teachers' interviews.

Confidence, flexibility, thoroughness, the ability to pick up new ideas quickly and with a minimum of teacher support, are all highly praised qualities. This suggests that, like the primary teachers, the secondary teachers felt positive towards those children who understood explanations readily and did not, therefore, seem to challenge their teaching ability.

All the qualities mentioned as positive by the teachers reflect the ideal pupil — one who is lively and interested (not too lively and therefore disruptive!), conscientious, little trouble in class, and produces a lot of work to a high standard with a good grounding in the basic areas of mathematics which can be capitalized on.

Poor children are seen to be of 'low ability', 'unable to grasp mathematical ideas', as 'unsure of themselves' and 'nervy', unable to remember what was learnt before in order to apply it to the present. They fail to make connections. Why the teachers should typify the children in these terms is a moot point. It is often thought that these children don't understand because they have language problems, social problems or behavioural problems. Essentially for the poor child the problem is seen as one which is not necessarily susceptible to change at school. All the poor children are considered to have behavioural problems, which get in the way of the things which children need to be good at mathematics, for example, an interest in and curiosity about things around them, perseverance and enthusiasm.

Looking at the characteristics specifically ascribed to boys and girls, good girls persevered, were sharp, had flair and confidence, all backed up by a solid primary school background. The boys, especially good boys, showed natural talent and worked hard, were confident, flexible, and took risks. They, too, had solid backgrounds. Poor children, if they were girls, were more likely to be considered lacking in confidence and anxious, rushing through work because of feelings of insecurity. They were also accused of both overemphasis in presentation and being disorganized. Poor boys were mostly seen as having behavioural problems — disrupting classes or hiding their inadequacies behind an arrogant front. They, too, lacked confidence, but were seen as more demanding of support, presumably because if it were not given they would be more likely to disrupt the class.

In conclusion, then, there is an important and complex relation between psychological and sociological theories, classroom practice, views of learning and evaluations of pupils, which has material effects on the way in which girls are seen to perform and how they see their own performance. Girls particularly are often placed in a double-bind because their success is taken to be produced in the wrong way. In this way we can begin to understand the relationships between the girls' good performance in the test results and their poor evaluation by the teachers. The important relation, then, is between what a teacher considers to be the cause of attainment and the basis of the social relations of the classroom. In addition, the differences between teachers, given their own background and practices, produce different approaches to teaching and learning, different results and different evaluations.

How and why do girls and boys operate in classrooms in ways that lead teachers to make these evaluations? In the next chapter we shall summarize our explorations in this area. How do children understand themselves and their peers? How are these understandings similar to and different from those of teachers? And how are they produced?

Chapter Five
Interaction and Social Relations in the Classroom

In the last chapter we considered the importance of the relation between theory, practice and teacher evaluation. We argued that the criteria through which good and poor performance are understood often lead to evaluations of girls' performances as not produced in the correct way. In considering why girls display characteristics which teachers read in the particular ways outlined, it is important to consider the children's understanding of themselves. In this chapter we shall demonstrate how girls struggle to maintain in themselves those very characteristics which lead to pejorative evaluations of their work by teachers.

Repertory grids

Primary school:
Repertory grids were chosen because they permit at the same time an exploration of a child's particularity and, by aggregating the grids, a method of looking at commonalities amongst groups sharing a common context. The grid itself consists of elements, which in this case were people, particularly those at school who were considered significant in either positive or negative ways, and constructs which were categories used by the participants to make sense of their world. The children were all asked to write about people they liked and people they disliked, and videotapes of classrooms sessions were analysed. Five main themes common to both schools emerged and they were treated as dichotomies. These were:

1. being nice/not being nice
2. being popular/unpopular
3. being clever/not clever
4. being annoying/not annoying
5. like me/unlike me

These five were supplemented by constructs having to do with being good at the subjects which occupied most of their school days: mathematics, English, art and sport,. Thus nine constructs were elicited and used for all the interviews. Each child was allowed to choose her/his own elements in a process known as triadic elicitation, whereby each of the ten elements was chosen according to their similarity to or difference from either the interviewee or each other. 'Self' had to be included and a construct which was 'ideal self' (someone you wished you were like) was opposed to 'negative self' (someone you're glad you're not). As Phillida Salmon has put it,

> The adaptability of grid techniques to individual situations also means that it can be used for the assessment of interpersonal relationships. The assumption of common areas of construing . . . is, perhaps, particularly relevant to the field of personal interaction where each individual involved must have some understanding of the others' subjective world if communication is to be effective (Salmon, 1976).,

girls and boys
The results were aggregated and analysed in various different ways. All the quotes are taken from the interviews conducted at the time of filling in the grids. Fourteen children comprised the original sample and the other sixteen were chosen, on the basis of their results on the mathematics test, as being good or poor at mathematics. The final total included eighteen girls and twelve boys, reflecting the proportions of the sexes in the classes of the two schools.

Looking at the choices of 'ideal self' most children chose others in their class. The boys most often chose other boys whom they considered good at sport, particularly football. All the boys who wished they were like famous people chose sportsmen: all except one was a footballer; the exception was a boxer.

Without exception the girls chose other girls within their class as people they wished they were like. Their reasons ranged from 'she's nice, pretty, kind' to one girl who was not good at mathematics wishing she was more like a girl who 'nearly everyone likes her, Miss, 'cos she can do her sums properly — sometimes she helps you'.

The choices of 'negative' self, i.e., someone the children were glad not to be like, showed an even narrower range of choices. Twenty-one out of the thirty children chose someone in their class. At both schools children chosen were all those who performed poorly in class, suggesting that the children's definitions of 'good' and 'poor' performance were situated

within the same nexus of relationships as those of the teachers. Two boys
and two girls chose the 'opposite sex' as people they were glad not to be.
Why were the girls glad not to be boys?

> . . . because they're dirty and spiteful . . . They're brutes . . . they're
> hard on girls sometimes.

RW: So you're glad you're a girl?

> Yeah, they're softer than boys. They're more intelligent in some
> things. Most things.

Another girl had a different reason:

> 'Cos if you buy clothes girls get more choice and things like this.

RW: There aren't things that boys have that you wish you had?

> Yeah . . . fun . . . yeah, I think they do (have more fun) 'cos they
> play football and games like that.

RW: Why don't you play football?

> It's not reasonable, none of the girls in my class plays football.

One of the ways in which these children make sense of their lives has
to do with the differences between the sexes. It was clear from the choice
of friends that there was little cross-sex friendship. No girl at either of
the schools chose a boy as someone they related to in any positive way.
Only one boy at school J2 chose a girl and his comments serve to underline
some of the themes we shall take up later. He said:

> Well as she's a girl, right, she just makes me laugh. And right, if
> I get into a fight or something she always tries to stick up for me.

Boys' reasons for disliking girls were not clearly articulated — unlike the
girls' reasons for disliking the boys — but an analysis of the interview
tapes shows two different worlds constituted by sex.
  For boys, sport is very important and defines each boy's place within
the class's relationships. Life is lived freely in the playground and other

spaces, and girls are mysterious and peripheral. For the girls the important people at school are nice, kind and helpful and there is little mention of any activities outside school except as they have to do with the family. There is a feeling of separateness between the two, which only come together with regard to another set of practices around pedagogy in the classroom.

When the grids were compared the most interesting data was concerned with the relationship of the construct 'clever'/'not clever' to the subjects which the children did in class. For boys, cleverness and being good at mathematics were close together. Girls linked cleverness and being good at mathematics with being good at English and being popular. This relates to the sets of practices observed in the classrooms (at both schools) and to the data from the interviews. Girls seem to link being good at their work with being a nice, kind and helpful person.

The following extracts are from the interview with Patricia, a girl in the sample considered to be poor at mathematics.

Well, I mean she's nice and all that 'cos I didn't know an answer and she goes 'What's the matter?' I go 'I'm stuck' she goes 'What number?' . . . and she told me the answer . . . She helps me with my maths.

So, people who are nice, etc., tend within the classroom to be those who are good at their work. They fit within two frameworks, one to do with being female, the other to do with accomplishing the work set in the classroom. This can be expressed in another of Patricia's statements about the opposite sort of girl — one who can't do her work.

She doesn't do nothing, that's why she wants to sit next to me so, she can copy all my work. I don't let her no more . . . I used to let her 'cos I didn't think . . . She got it all wrong 'cos I got it all wrong.

This shows how children understand the teacher's pedagogical frameworks about learning by doing things for oneself, and are prepared to apply them to themselves and others.

'Nice', 'kind' and 'helpful' are seen as feminine characteristics. Cleverness is associated with these, but helpfulness seems to be the most important and can be substituted if a girl is not understood as clever.

being good and poor at mathematics
When the consensus grids for all the children (boys and girls) chosen as
good and poor were analysed it seemed that, for all children, being good
at mathematics was related to being clever. Poor children of both sexes
tend to see mathematics in a cluster with other unanticipated constructs.
For example, two of the boys from J2 who were poor at mathematics
linked being good at mathematics to the construct for annoying. One of
them said:

> . . . sometimes I find the sums quite long and that, and I can't work
> 'em out so I just leave 'em . . . I am bothered about maths but if
> it comes to a hard sum and I can't do it I leave it and go on to the
> rest, but when I've finished if I can do it I do it, but if I can't I
> just leave it out.

Several of the girls (3) at this school linked 'annoying' with 'being good
at games'. When a consensus grid was produced for the schools the one
for J2 showed that there was a close relationship at this school between
'self' and 'being good at games', suggesting that it was important at this
school in the way children were positioned. At J1 being clever was close
to the constructs about the subjects, suggesting that at this school there
was more of an emphasis on class work.

First year secondary:
The techniques described above (p.64-5) for eliciting constructs and
elements were used with this group of children. Eight constructs were
chosen which seemed to encompass most of the choices made by children.
Of these eight, five were to do with the most often mentioned subjects,
namely, sport, science, French, English and mathematics. The remaining
three were 'personal' constructs, being concerned with personal attributes
— positive and negative. They were, 'friendly and understanding', 'boring', and 'clever'.

Thirty-two children from five tutor sets (a third of the total number
in the five sets observed) were chosen for interview: eighteen girls and fourteen boys, reflecting the proportions in the first year. They were also
divided by whether they were good or poor at mathematics (apart from
the original sample children this was based on their test results) and by
the numbers and proportions in their respective sets. In general, a third
of the tutor sets were interviewed.

When the grids were completed and analysed it was decided to concentrate on an analysis of consensus grids as showing differences or similarities between groups. From these we could see the relationship between the different groups not only on the constructs offered, but also at the particular time they were administered. The latter is very important, providing a snapshot of how certain children felt at a certain time, and about specific events and people. They cannot be taken to be universally applicable.

girls and boys
A comparison of the boys' and girls' grids reveals differences between the perceptions of self. Boys rated themselves more highly than girls on the constructs 'sport', 'science' and 'friendly and understanding'. They rated themselves as less boring than the girls. The importance of sport to boys is reflected in their ranking of their ideal selves as better at sport than they actually were. Girls did not rank more highly than the boys on any constructs. This would suggest that boys tend to see themselves more positively than girls, especially on the constructs provided here.

An analysis of the ways in which elements were chosen reveals which qualities are considered positive and which negative by this group of children. Each child was asked to select an ideal self — someone they admired or wished they were like in order to provide us with some notion of the sort of people considered significant. Eighteen/thirty-two chose people they knew as their ideal selves — twelve of them chose classmates (seven girls and five boys). The girls' reasons were diverse: 'She makes friends easier than I do' or, as one of our sample children, Ruth, said, 'She's just like a boy but she don't wear trousers. I'd like to be like that. She's fun.' Another of our sample, Patricia, put it, '. . . she's so clever she understands the cards (SMILE) straight away'. Another chose a girl for a variety of reasons. '. . . a nice girl . . . a nice name. She's nice mannered, she has nice clothes and she's helpful, the very opposite to me.'

This latter comment suggests that this girl is aware of what an ideal girl should be and her list of this girl's qualities is a list of female stereotypes which she cannot attain because, as she puts it, quite sanguinely '. . . it's not in my nature'.

Girls tended to choose classmates whom they admired for their personal qualities, whereas boys chose classmates whose achievements or performance in school activities surpassed their own. As George put it about his choice, '(he's) good at a lot of other subjects at which I'm not'. Or it was 'just personality', or because the boy was 'good at nearly every sport and he's clever', — an enviable combination! None of the boys chose a woman as their ideal.

Whilst none of the girls chose boys, two did choose men — mainly because of their occupations. One was a doctor — this girl's ambition, the other was a male mathematics teacher 'because he's clever'. Obviously there are no similarly highly valued female characteristics.

Only boys chose famous people as those they aspired to. They were men who excelled in field in which these boys were particularly interested — a judo expert, a scientist and a guitarist — and related to important outside interests. Girls, on the other hand, tended to pick people who possessed characteristics which they valued. This suggests that girls' out of school activities are either less important to them or, as indicated by the primary data, likely to be non-existent. Boys and girls in the first year of the secondary school are still leading quite separate lives, with quite separate senses of what is important. For the girls it still implies a more circumscribed way of life. Boys are able to range wider and have wider horizons. The split between the private and the public, femininity and masculinity, is evident here.

On the other hand there seems little envy or wish on the part of girls to be more like boys. Even Ruth's comment earlier about her friend being '. . . like a boy but she don't wear trousers' suggests that girls may want to be like boys in some respects but not all.

If we examine the choices made for 'negative' selves, i.e. those people one was glad not to be like, fourteen out of eighteen girls chose people they knew and five of those chose boys. Reasons ranged from their meanness, bullying and showing off, to causing trouble. Six out of eighteen girls mentioned how silly and irritating the boys' behaviour was and other girls mentioned their stupidity and selfishness. But, in fact, all these negative personality traits were also used in reference to girls as well. Five out of seven boys also chose a classmate as someone they were glad not to be like, but only one chose a girl as 'bossy'. The rest were all boys and the reasons were similar to those of the girls — silliness and meanness. Half of the boys (7/14) mentioned girls as possessing those negative qualities which they most disliked. These ranged from blaming the boys and getting them into trouble for bullying, to being boring, disruptive and noisy.

'Niceness' was the positive characteristic most often mentioned when elements were being chosen (16/32). When asked to explain further what it mean to them, ten children said it was to do with being friendly and understanding, the rest saw it as being popular and getting on well with others, or related it to cleverness. Nice is a catch-all term, most often used by and for girls, which embraces a wide range of attributes from looking good, having good manners to being able to get on well with others. Generally, it embraces a set of female stereotypes and is not often used to apply to boys. The other two most frequently chosen characteristics were also exclusively applied to one sex or the other. Only girls chose helpfulness after niceness and boys chose prowess at sport. Other characteristics exclusively chosen by girls were quietness, responsibility and funniness. Only boys chose being good academically.

In terms of negative qualities, the most common choice of both girls and boys was being 'flash' (showing off in class) or being boring. Bullying, mucking about, were chosen by twice as many boys as girls as a negative. Girls had a greater variety of negative characteristics, including being soppy (this in relation to boys), meanness, telling tales, not being clever, being shy, or rude, or selfish.

Only two girls mentioned bullying directly, but another mentioned being mercilessly and hurtfully teased by boys, which suggested that maybe the same behaviour was called different things. If the two categories were taken together, twice as many girls as boys mentioned that sort of behaviour.

As in the primary interviews cross-sex choices were rare in any positive sense. Boys and girls chose the opposite sex as negative elements. It was only in one tutor set (3) that boys and girls chose members of the opposite sex as someone whom they liked or would have liked to be like. In this set the most positive and popular person chosen was a girl, whose combination of cleverness and prowess at sport made her universally popular. Otherwise, the boys and girls seemed to co-exist in separate worlds.

However, when the children's individual grids were correlated to the grids for their tutor set and then for their sex (regardless of performance at mathematics), twenty-six out of thirty-two of the children interviewed had higher and therefore more positive correlations with the consensus for their set than for their sex. This suggests that, at this stage in the first year, children tend to feel less differentiated by sex than by the tutor set in which they have been put and in which they spend most of their time. The solidarity thus shown could have been reinforced by the newness of everything else in the school and by the particular position of the first years as the lowest of the low, the smallest fry in the pan.

being good and poor at mathematics
The comparison between the grids for those girls who are good at mathematics and the boys who are similarly good shows an interesting

divergence, with the boys ranking lower than the girls on prowess at mathematics. Good boys don't think they are as good at mathematics as the good girls. Looking at the rankings of the girls' and boys' ideal selves, it is possible to summarize the areas which the sexes felt were most important in that they gave them high ratings. Girls' ideal selves rated much more highly than both the girls and boys on the constructs for 'friendly and understanding' and for science, whereas boys, inevitably, came higher for sport.

The only other set of comparisons between the grids which produced any different results was the comparison between poor and good boys. Poor boys rated themselves more highly for the constructs for French and for mathematics than did the good boys. The similar comparison between girls showed the poor girls as rating much lower than the good girls on all the constructs. This suggests that prowess at mathematics is not important to boys or that it is not as important as it is to girls. It also suggests that ability at mathematics is not considered as important to their self-perception by boys.

Children who were poor at mathematics, both boys and girls, when compared with those of their sex who were good at mathematics came off worse in the comparisons. Poor children seemed, regardless of sex, to have much lower opinions of their own worth in relation to these particular constructs than did good children, regardless of sex.

Fourth-year secondary:
To generate the sample for the fourth-year secondary work a different strategy was invoked. The first-year sample had been chosen by their primary teachers as good and poor and their progress through the educational system was followed. In looking at the fourth-year we had to revert to the tactic employed originally, which was to ask their mathematics teacher — who was also their form tutor and had been since their first year — to select those children, boys and girls, whom he considered to be good and poor at mathematics. Eight children were chosen — four girls, two good at mathematics and two poor, and four boys similarly divided. One of the poor boys had to go into hopsital in the middle of the field work and so was excluded from the RG interviews. As only one fourth-year class had been observed because of the time constraints it was decided to make up the rest of the sample of children to be given a grid with other children in the class considered significant. This was determined by the amount of time the teacher spent with them, the number of times

they were mentioned by other children when interviewed and their appearance on the videotapes. As a consequence five other children were chosen to be interviewed and, in fact, they all turned out to be children considered poor by their teacher — three girls and two boys.

These particular children were also considered significant in that their performance in the English lessons observed shared differences in the styles adopted in the different lessons.

In total twelve children were interviewed, all of whom did mathematics together and most of whom did English together — seven girls and five boys. The constructs were again supplied by us after careful observation and after reading comments written by the tutor sets for their tutor on their thoughts about their lessons. By the fourth-year optional subjects had been chosen, so that not all the pupils did the same timetable. We decided, therefore, that the only academic constructs which could be supplied would be in the three 'core' subjects — mathematics, English and Social Education. All these subjects are compulsory and we could be sure everyone would have experienced them. The other four constructs — 'personal' ones to do with the traits which were considered important by the class — were 'helpful', 'hardworking', 'boring' and 'like best'/'like least'.

girls and boys
Looking at the comparisons between the consensus grids — and the same comparisons were made between sub-sets of pupils as for the first years — it is clear that by the fourth year there is very little common ground between the sexes. The correlation between the boys' and girls' grids is very low (0.47). The construct with the lowest correlation between the groups is that for mathematics, and the highest correlation is for helpfulness. The girls rate mathematics more highly than do the boys.

Turning to the interviews, and the choices made of ideal and negative self, all girls chose a class mate as their ideal self, except for one girl who chose her mother. No girl chose a male ideal and no two girls chose the same ideal, suggesting that as differentiation in subjects increases as the children get older so, too, does any consensus about admired figures. Maybe the importance of role models in the imitative sense diminishes with age and experience, and with the importance of individual choice. The reasons for the choices of ideal self were mainly to do with academic prowess — except for one girl who chose someone who was better looking than she was ('That's the only thing I really worry about'). All the other girls mentioned the fact that their ideals were clever: '. . . she knows everything'; 'She's good at maths.' Two of the girls were more specific: 'She's really clever, she can really get down to work, whereas, if I had to be top of the class, I'd have to be really pushed . . . and I'd have to work really hard. In this school it's quite hard 'cos everyone's always mucking around and there's so much talk, but she can just work in any atmosphere and just get on.' As another said about a girl whom she'd chosen as her ideal, 'She's just a really nice person and she's clever. Everything. She's got everything.'

In the choices made by the boys of their ideal selves all were male. Three were friends with whom they had lessons, one was a fictional character and one was a male teacher. The reasons given were similar to the girls, in most cases and revolved around being helped when in need, and people who were nice and clever: 'He's just a really nice guy, really clever and that's it'; '. . . because he's good at maths'; '. . . good friend, nice, and when you need help he's there'. One boy had a specific reason for choosing his ideal: '. . . he possesses some qualities of . . . guts which I don't'. The boys in this sample were more likely to refer to the importance of being helped than were the girls. As for negative selves, all the girls chose someone with whom they had lessons, but this time three of the seven choices were boys, although not the same ones. The girls found it more difficult to articulate their reasons for disliking the boys: 'He just gets on my nerves' or 'He really makes people's lives a misery . . . he's just so horrible. I'd hate to be horrible like that.' Two of the girls chose the same girl as someone they were glad not to be like, but not for the same reasons. One of the girls felt she was lively and interesting but talked too much, which meant that she didn't get a lot of work done. The other girl knew more clearly why she was glad not to be like this girl: 'She's not particularly bright . . . she's not really stupid or anything but she just seems to be. I don't like the way she dresses . . . she hasn't really got a best friend and she's always, like, hovering around with two or three people or just by herself and I'd hate to be like that.' This girl is different, doesn't fit in and is, therefore, an outsider.

Looking at the choices of ideal and negative selves, one girl was chosen by two people — once as an ideal and once a someone they were glad not to be like. The differences reflected the two aspects which have been coming out as important to these girls' choices: the fit or lack of fit between being good at work and being a nice, sociable person. The two often seem in opposition and where they are together there lies perfection! The choice of this girl as an ideal related to her ability at mathematics and ability to help, and her choice as a negative self reflected her personality: 'She's not sociable. She doesn't talk to people; she's too shy.'

Two girls, one good at mathematics and the other poor, chose each other as ideal and negative selves. Not surprisingly the poor girl wanted to be like the good girl whom she admired and who helped her, whereas the good felt much more negative about the poor girl. Despite claiming her as a friend, she was clear about her reasons for not wishing to be like her. '. . . she's so unlucky. She tries so hard . . . she gets nowhere. She's a very stereotyped person and I really can't change her.'

As for the choices of other elements, girls were more likely to choose male elements as people they felt positive about, or who were like them in some way, than boys were to choose female elements, except for one sister or a mother. This suggests that it is easier for girls to see positive things about males than it is for boys about females. It also reflects what the repertory grid consensus grids pointed up, which is the separation between the groups of children. Their experiences, reflected in different curriculum choices, give them different perspectives and there seems to be little overlap or any wish for overlap. It is telling that it is possible for girls to wish to be like boys whereas the opposite choice seems less likely. The boys' negative selves were all male and all in the school — two of them were the same as those chosen by the girls and for similar reasons: talking too much and showing off.

The quality which was chosen as the most positive by all the children was the ability to get on well with others and share common interests. The next quality which was chosen by most of the children, regardless of sex, was the importance of being a friendly listener. A quality chosen by all the girls and a couple of the boys (those not good at mathematics) was helpfulness. 'Cleverness' was considered a positive quality by five out of seven of the girls and three out of five of the boys, i.e. by two-thirds of the children. 'Lively' and 'nice' were mentioned and so were the importance of confidence and perseverance in succeeding at school. Being 'interesting' (with no more explanation) was a positive quality mentioned only by boys in their choices of elements, and 'hardworking' (as a positive quality) was only mentioned by girls.

The negative quality often mentioned (four girls and two boys) was insensitivity in terms of being unaware of others i.e. not listening or being unsympathetic. Showing off and being irritating also annoyed most people. Girls were most likely to mention shyness as a negative characteristic. Generally those characteristics most often mentioned were those that would set children aside from their peers: unhelpful, talking too much and therefore likely to incur the wrath of the teacher, unsociable, odd. Interestingly, qualities which might be supposed to inspire sympathy like shyness, unsociability, unluckiness, looking funny, were perceived as negative characteristics and engendered dislike rather than sympathetc understanding. Of other characteristics 'selfishness' was mentioned only by girls and 'boring' and 'unintelligent' only by boys.

being good and poor at mathematics
A comparison between the sub-sets of children who were good and those who were poor at mathematics shows that there is here, too, a low correlation between the grids. The poor children's rankings of themselves show that they considered themselves inferior to the children who were good at mathematics, both in terms of their performance at mathematics and in social education, and they saw themselves as much less hardworking.

If we compare poor girls with good girls, the poor girls ranked themselves lower on all the constructs except 'helpfulness'. This suggests that 'helpfulness' was all that they had left as a positive concept if they were not good academically and did not consider themselves hardworking. Yet it does not seem to be seen as a very positive asset when the rankings for ideal self are considered, and the poor girls ranked their ideal selves much lower than the good girls, and the good girls' ideal selves, on helpfulness. This is also true of the comparisons between the good and poor boys, who ranked themselves more highly than the good boys only on helpfulness and lower for hardworking, mathematics, etc.; but whose ideal selves ranked higher than the good boys on hardworking and mathematics and lower on helpfulness. Helpfulness, like niceness, is a catch-all for people about whom one can think of nothing more positive to say. It has positive aspects, but overall it seems not to be a highly valued attribute.

One of the most interesting things about the consensus comparisons of the fourth-year children is the differences between the different groups. In the first year the differences between groups of boys and girls, and children who were good and poor at certain subjects were still small but the intervening three years serve to widen the gap. Option choices and possibilities for taking either of the public examinations must reinforce the split between the children considered good or able and those not.

2. Mathematics in particular
All three age groups were specifically asked about their feelings towards mathematics.

Primary school

At J1 out of fourteen children nine gave positive responses, three disliked it and two were non committal. At J2 the sixteen children divided evenly between those who liked it and those who didn't. When we look at the distribution by sex we find that eleven out of the eighteen girls (61 per cent) expressed a positive view about maths. Only six out of the twelve boys did so. Why did these children like mathematics? At J1 most seemed to do so because they were good at it. One boy was more specific: 'Sometimes I'm good, sometimes I'm not. I'm quite good at doing the numbers, adding up and taking them away. I don't like symmetry and grids 'cos sometimes I get it wrong . . . and that's annoying.' Whereas a girl at J1 said 'Well, sometimes I understand it and sometimes I forget what I'm doing and think about other maths I'd done what are easier . . . the best maths I like doing is making and finding out shapes.' Another J1 girls said: 'It's interesting when you don't know how to and when you try to find out, and it's nice when you know how as well.' At J2 all the negative replies about mathematics, from both girls and boys, had to do with how hard and boring the subject was: 'All that adding up and taking away, timesing and division. Boring' (boy); 'I don't like it, the — er — doing graphs and you have to do with sand and buckets and everything' (girl). A girl chosen as not very good and whose test results bore out this judgement had a clear idea of how and why mathematics was important. 'Maths is good 'cos it helps you. Well, let's say can I have ½ lb of tomatoes and you give them a pound. All you have to do is to halve it . . . It's rather easy. If you don't know your maths that's the end of you.'

First-year secondary

The replies from the first-year secondary children were more informative. With regard to work the children were asked about mathematics specifically and also about the mathematics scheme used — SMILE. Nineteen/twenty-three children, or 59 per cent, liked mathematics: broken down by sex this was made up of 61 per cent of girls and 57 per cent of boys. The most positive sub-set of children were the good boys of whom six/seven claimed to like mathematics. Interestingly 60 per cent of this same group of boys disliked SMILE. It was overwhelmingly the good children of both sexes who liked mathematics (85 per cent). A majority of poor girls also claimed to like it, but only 28 per cent of the poor boys did so (2/7). The reasons given for liking mathematics ranged from finding it useful and interesting to liking it because one was good at it and it was easy — the latter reason from a boy. Most girls who liked it did so because it was interesting. Of the thirteen children who said they didn't like it seven were girls. Their reasons fell mainly into two categories — either they felt they couldn't do it and were no good at it, or they didn't like the subject or the teacher. Six of the boys claimed not to like mathematics but were still prepared to see themselves as quite good at it (as the consensus grids showed earlier poor boys rated themselves as better at mathematics than the good boys despite their performances). Three girls and two boys — all poor children — claimed to know they were no good at mathematics.

When asked about SMILE there was a more positive response, 22/23 children saying they liked it. If we compare the figures it seems that the first-year pupils do not see mathematics and SMILE as synonymous. SMILE certainly evokes more positive responses than mathematics except for good boys. Broken down by the sub-sets the figures for mathematics and SMILE are as follows:

89% (8) good girls	liked SMILE compared with 66% (6) who liked mathematics
67% (6) poor girls	liked SMILE compared with 55% (5) who liked mathematics
43% (3) good boys	liked SMILE compared with 85% (6) who liked mathematics
71% (7) poor boys	liked SMILE compared with 28% (2) who liked mathematics

So this surprising set of figures shows that good boys who felt overwhelmingly positive about mathematics feel much less enthusiastic about SMILE. Their dislikes centred on the boredom of doing the same or similar work. They felt it was childish, like being back at primary school, and that it was as easy as primary school work. One of the poor boys disliked what others found challenging, the variety of the workcards. He would have preferred to work through a book and thus to have seen more tangibly his achievement. All the boys who disliked SMILE would have preferred to do other kinds of work, either from the blackboard as a class lesson, or to be allowed to work in groups. Certainly exploring the use of other teaching strategies may help to keep interested those children already feeling disaffected in their first year. Reasons for liking SMILE, on the other hand, ranged from 'interesting', 'enjoyable', 'easier than other maths'.

The most interesting difference seems to be that the group who felt least happy at their new school (the good girls) felt happiest about doing work in what was to them a familiar form. Whereas the boys who seemed to welcome the changes brought about at secondary school disliked SMILE precisely because it put them in mind of their primary school, a stage they felt they'd left behind.

Fourth-year secondary
The fourth years when asked specifically about mathematics — the teacher and the subject — claimed mostly to like neither. Two of the poor boys claimed it to be 'boring', their most common complaint, followed by one boy's declaration that 'I'm no good'. Five/seven girls claimed not to like the subject. Two said they just weren't very good and one claimed to be 'not very mathematical'. Interestingly, this latter girl was the only one to mention SMILE by name, saying she preferred it: 'Yeah I prefer to do that, when we're all doing different topics, than when we're all doing the same thing as a class. You can go at your own speed.' One girl claimed to dislike mathematics because of the teacher, but generally the girls and boys who disliked the subject attributed it to their own inabilities. As one boy said 'I could never be good at maths, even if I tried.'

One girl expressed her dilemma clearly. She had left her primary school as a band-one child on all three comparability tests: mathematics, English and verbal reasoning. And this, even though she claimed she had 'absolutely detested' mathematics at her primary school. Once at secondary school things changed: 'I don't know, at some point I thought, oh, I'm no good at maths and I didn't really try I suppose.' She recognized what had happened to her and now that she was reaching a crucial stage in her career was beginning to want to get down to more work, but it wasn't easy: 'Well, now I'm trying to work again. But also I get easily distracted by people . . . who're good at maths anyway, so they can distract us and get back on with their work again.' Why did she lose interest? The subject was boring and she thought her teacher could be callous and hurtful to those he did not like and who needed a lot of help. It also reflects the concern voiced earlier about the importance of keeping control, so there was an atmosphere conducive to work created in the classroom. Other girls were less clear about their reasons and just thought that they were no good and that was it.

Mostly it was the children considered poor at mathematics (except one of the boys) who claimed to dislike it. They often attributed their dislike and lack of success to themselves not being very good at it, but it seemed also that that was how their teachers treated them and they were stuck in a circle of performance which could not be broken. They read the teacher and the lessons as boring and uninteresting and he felt them to be uninterested and irrecoverable, and so they were.

Summary
Let us summarize briefly our analysis of this section, setting out what we think are the most important conclusions which can be drawn.

In the primary school it is significant that certain characteristics are stated as important by the girls. They are valued in people whom they know; they cohere around such terms as 'nice', 'kind', 'pretty', 'helpful'. Boys, on the other hand, are much more likely to desire to be like famous sportsmen, not people they know, thus setting distinctions between what, in common-sense terms, is called 'personality' and 'achievement'. The girls appeared to favour those feminine qualities which Angela McRobbie (1978) described as being those necessary for 'getting and keeping a man'. The boys' qualities emphasize their public presence. However, the latter are far less oriented to academic attainment than the girls. The important link for girls is between being good at mathematics, clever and helpful, nice and kind. So, clever girls are valued if they help others with their work. Poor girls, if they can't be clever, can at least be nice! In other words good girls strive to be liked by other girls. To do this they must display certain characteristics of femininity. Yet is is precisely those characteristics which lead the teachers to make evaluations of hardworking, conscientious, passive, and therefore not displaying the outgoing, divergent, active characteristics associated with 'proper learning'.

The girls, then, have more orientation towards formal/academic work than the boys, whose 'macho masculinity' in terms of an anti-intellectual liking for sport (see Willis, 1977) is already in play. However, the girls' orientation is, by itself, not enough.

By the first year of the secondary school boys still rate sport as important. Again, the girls relate to qualities in their classmates considered good, whereas boys choose achievements they would like for themselves. Boys rate their own achievements in terms of their evaluation of their own performance higher than do girls. As in the primary school, the distinctions between public and private, achievement and personality, remain.

Girls and boys here, as in the primary school, are still very separate groups and they tend to recognize in the other group bad qualities — that is, boys invest them in girls and vice versa. Girls still rate attainment in mathematics as more important than the anti-intellectual boys. However, the 'best' children are those who are located powerfully in respect of all those practices valued by the children. That is, a child who is clever, good at sport and popular manages successfully to attain positive valuation by peers in all areas.

By the fourth year of the secondary school girls still value mathematics and being clever more highly than boys. Again, for girls to be nice *and* clever is the aim. Boys lay more stress on being helped than before. However, those qualities valued most appear to relate to the ability to handle social relations effectively: to be shy and boring is regarded as bad. *In relation specifically to mathematics,* while girls are more positive than boys about the subject in the primary school, by the secondary school important differences occur. While good girls like SMILE work, good boys dislike it, or find it boring. The reasons why girls like it relate to its safety and continuity with the practices of the primary school. The safety is precisely what allows them to continue in the hard work/feminine/helpful constellation, and precisely what leads to pejorative evaluations by the teachers. As we shall see it is the reason why teachers prefer girls to take CSE rather than O level. CSE is school-based, more importantly it is course-work based, it is safer, requiring less pushing. The circularity with which practices, teachers' evaluations and the self evaluation of the girls fit together is extremely important. The first-year boys want to leave the practices of the primary school behind and dislike SMILE for that reason. This, of course, allows for the development of a different participation in the classroom which is, in turn, evaluated more positively.

By the fourth year most children, however, do not like mathematics at all. They find it boring. In turn the teacher feels they are uninterested and is disheartened. Bored children do not make an effort and, therefore, do not do well. There is again an important circularity or circulation between evaluation, practice and performance: no simple analysis of cause and effect will sufficiently explain this complexity.

Importantly for our analysis of girls, the relationship between the promotion of femininity, success at mathematics and its evaluation by the teacher is a complex one. Femininity is not to be understood as in *opposition* to academic success, as has been suggested in some approaches to gender using stereotyping which see femininity and passivity as the qualities which help produce girls' failure. Our analysis suggests that the matter is not so simple. Femininity is not antithetical to success, though what is understood as feminine behaviour is also understood as antithetical to 'real learning'. The difference between this analysis and one which uses stereotyping is that the latter would advocate freeing girls from femininity, making them more like boys in order to succeed. But since some girls *do* succeed anyway, the matter is not that clear cut. We do not, therefore, in any simple sense advocate encouraging girls to be more like boys. Rather, our analysis points to problems both in the pejorative way in which their performance is evaluated *despite* success, and to those practices which help maintain both femininity and masculinity.

In the next chapter we shall examine how the classroom practices themselves help promote and produce success and failure in girls and boys.

Chapter Six
The Classroom

We have argued in previous chapters that there is a particular combination of classroom practices and an understanding of mathematical learning which produces failure in girls, and that in consequence girls are put into the position of being successful but not succeeding. Here we shall examine some of the parameters of the classroom production of this situation for girls, illustrating with examples from our own study. Since the amount of data, case material and observation is vast, we can provide only a glimpse of the kind of analysis which we have carried out. We have decided that the most effective way to present it is to set out some of our more important analytic categories and then to illustrate these with reference to the case of particular children (girls and boys) who represent specific polarities in the positionings we describe.

In both primary and secondary classrooms our fieldwork consisted of fieldnotes, observation of classroom practices and videotapes of the specific performance and interaction of the children in our sample. Each of these children was videotaped for one hour. Together, the fieldnotes, test, interview and grid data, and videotapes, form a massive amount of information on children and classrooms. Our discussion here will concentrate on the presentation of fragments from our transcripts illustrative of particular concepts derived from our analytic framework. We shall begin by elaborating what these are.

Our pilot work (Walden (Eynard) and Walkerdine 1981, 1982) identified certain concerns which we felt offered potential explanatory devices for the apparent phenomenon of discontinuity. As we have said, we have not discovered a simple discontinuity as such, rather continuities in the data which support our previous work on early success, and make of later performance much more of a problem than was previously envisaged.

Let us set out in detail some of the categories we used and have continued to develop in relation to this data.

Power, positioning and gender

We have argued strongly against an analysis which understands girls as powerless because feminine and against a model of girls simply 'squeezed out' of academic performance. Relations of dominance and subordination, and of power and resistance, can be explored in relation to the social relations of the classroom. We shall argue that girls are not subordinated in any simple or once-and-for-all way but that they can move from powerful to powerless positions from one moment to the next. Femininity and academic achievement are in this analysis not incompatible, but their relationship, as we have been trying to show, is neither without *problems* nor without specific *effects*.

Our analysis depends upon a theoretical framework elaborated elsewhere (for example, Walkerdine 1981; Henriques et al, 1984). However, for our purposes here it is important to state that we are critical of a monolithic view of power which understands it as a unitary possession vested in particular individuals simply by virtue of their institutional location. By such analyses power is invested in teachers and boys and not in girls. While we would not wish to argue that the relations in terms of institutional position of teacher and pupil, boy and girl, are ones of simple *equality,* what we are addressing are the ways in which relations of power and powerlessness are produced in classrooms and the forms and content of understandings available to teachers which operate in educational practices. In their formal sense these can be described as power and knowledge relations (see for example, Henriques et al 1984). Such an analysis implies criticism of some formulations of the sociology of knowledge (for example Young, 1971) in which knowledge is powerful because it is *possessed* by a particular group (such as teachers).

A particular site or positioning which, as our pilot work demonstrated, allows girls to be powerful in the primary classrooms is the position of *sub-teacher.* By being positioned like the teacher and sharing her authority, girls are enabled to be both *feminine* and *clever;* it gives them considerable kudos and helps their attainment. In a variety of ways the *relations* of power and powerlessness, helping and being helped, may be shown as existing between teachers and children and between children. Some girls will be helped by one set of children and be helpers to another, powerful in one set of relations, powerless in another. For example, a girl may be popular and not academically good. By examining those practices and contexts which we outlined in the last chapter we can see that there are relative (and to some extent cumulative) powers. A girl who is located as powerful in helping and in sport, for example, has a very high status with other

children. However, she may still not be considered to have 'flair', to be
'really' or 'naturally' bright by the teacher. There is, then, a difference
between the pupils' and teacher's estimations, and, this in turn relates to
the issue of femininity and classroom attainment. They may be com-
plementary or even contradictory. These different positionings produce
and affect different girls differently, but their total effect helps produce
the possibility and reading of attainment in the classroom.

Although the positioning of mathematical success in terms of *rule
following and challenging* interacts with the masculinity and femininity
dimensions we have argued against a simple reading of the production
of independence and autonomy in girls by making them more like boys
as a prerequisite to academic success. What we shall explore is the *effect*
of a certain kind of confidence in rule-challenging procedures in the evaluation of performance by the teacher. We argued in our early work on
primary school mathematics (Corran and Walkerdine, 1981) that the practices of mathematics teaching relate to procedural rules. These rules are
both behavioural and organizational in relation to the classroom, and
internal to the organization of mathematical knowledge itself. In order
to be successful children must follow the procedural rules. However, breaking set is perceived by teachers as the challenging of procedural rules
internal to the organization of mathematical knowledge itself. It is read
as 'natural flair'. In the first instance naughtiness (most often displayed
in boys), i.e. breaking behavioural rules, can be taken as evidence of a
willingness to break set, to be divergent. Consequently girls' good
behaviour is taken to be evidence of passivity, rule-following, and hard
work. Later, however, bad behaviour in class tends to be expressed as
anti-intellectualism which can no longer be read as playful but as opposi-
tional (Walkerdine, 1981; Willis, 1977).

To challenge the rules of mathematical discourse is to challenge the
authority of the teacher in a way which is sanctioned. Rule following and
rule breaking are both received forms of behaviour even though they are
antithetical to each other. If there are considerable pressures specifically
on girls to behave well and responsibly, and to work hard, it may well
prove more than they can bear to break rules. Firstly, they would risk
exclusion by others for naughtiness and secondly they would require the
confidence to challenge the teacher. Such contradictions place them in a
difficult, if not impossible, position. Our analysis leads to the conclusion
that there are problems for girls, but *not* of the order of a fundamental
or essential lack of ability, or of missing out a natural sequence of correct
development towards cognitive maturity. Rather, social and psychic

relations coalesce to produce possibilities, positions and constraints which both allow and prevent certain forms of behaviour and of thinking.

For example, to understand the contradictions involved in rule breaking and the problems attached to speaking out (see Spender and Sarah, 1980) is very different from an analysis which suggests girls have simply either 'got something missing' or have been forced out i.e. by a patriarchal conspiracy, currently the most favoured forms of explanation.

In the fragments of case material which follow we shall present examples of the kinds of positionings which we have outlined.

Individual cases: primary data

We shall begin by focusing on Patricia — a girl in school J1 chosen by her primary teacher as poor at mathematics. We shall see how she is produced and maintained as helpless. She was one of a group who were considered all to be 'much of a muchness'. She herself chose as her ideal self a girl whom she considered, '. . . just like me really, not very good at things'.

The transcripts show her, along with her classmates, involved in doing an exercise on numbers. They have been asked to add all the numbers from 1 to 5, then 1 to 10, 1 to 15, and so on. Consonant with the practices in the primary school the teacher never makes explicit what it is he wants them to do. First they do some examples, then if it is not obvious to them he will explain the proposition he wishes them to understand. Eventually he talks of triangular numbers which is what the exercises have been leading to, but at this point the class are being given practice at manipulating numbers.

In Patricia's case it is apparent that she has misunderstood the teacher's instructions to add all the numbers. Her friends have to help her and she becomes steadily more anxious throughout the tapes. At the beginning of the second tape she tries to get the teacher's attention to see whether or not she is doing the correct thing:

What do I have to add it to, Sir? (to Jo her friend) He's just shown me how to do it, I'll find out, I've gotta add it haven't I? Sir, is this right? I've got it I think . . . Here you are, look here, here's the answer I got, Sir? Sir? Is this right?'

The teacher ignores her to talk to another child. Patricia continues to try and attract his attention:

Sir? Is this right? I'll go and have a look 'cos if . . . if he's showing
us how to do it I'll go and have a look . . . Sir, is this right? . . .
Sir, I can't do these.

The teacher's response, on looking at her work is to suggest that the work
is too difficult and that she should do some easier work, which has been
written on the board and explicitly marked as easy. This is an attempt
on his part to help her — work she can do may not leave her feeling as
demoralized as she obviously is at present. But all it succeeds in doing
is confirming for Patricia her position as not very good, because it offers
her security in her helplessness. Her position as powerless is underlined
by each of the other dimensions mentioned earlier. At her school the
teacher felt unable to intervene with Patricia other than to give her more
practice on lower-level work which he hoped would help. Patricia's
problem in attracting his attention meant she had to rely on her friends
for help. The following exchange is typical of Patricia's contact with her
friends. Patricia constantly requests her neighbours Ann and Jo to help:

J: (to Ann) She's done it wrong
A: Who?
J: Patricia
P: Well, how do you do it? . . .
A: You're still doing it wrong.
P: (to Jo) Am I doing it wrong?
J: Yeah.
P: Why?
A: 'Cos you're not copying the board.
J: You're not meant to copy the board. You're meant to work it
out for yourself.
A: You're not.
J: You're meant to work it out yourself . . .
P: One to three equals four, right? (she looks at Jo's book), Yeah,
Oh God, talk about thick. (she rubs out her work).
J: You're doing multiplication. It's add.

In fact Patricia has been adding the numbers but has failed to understand
that the task is to add *all* the numbers from one to three not to just add
one and three. In this episode Jo establishes her superiority in several ways:
by talking about Patricia to a third party whilst ignoring her; she offers
no advice on how to correct the problem, merely stating that she knows

what to do whereas Patricia obviously does not. Ann's response to Patricia is also dismissed since the children 'know' that within the pedagogic framework of their lessons learning is achieved by doing the work not by copying it. Jo is able to dismiss Ann and deliver the *coup de grâce* at the end by nonchalantly 'spotting' Patricia's problem. From this point on in the tape Patricia asks Jo for help eight times.

Patricia is, therefore, always put in a position of being the one who is helped. Jo is dominant in the exchange and understands both the immediate problem and the rules of the classroom and acts as a *sub-teacher to Patricia* (and to Ann).

Let us take an example of a good girl in the primary school. Elizabeth is in a different position. At the same primary school as Patricia (J1) she was considered good by the teacher: 'She's interested and she finds it fascinating and she's interested in the odd bits of it.'

She liked mathematics when asked and considered herself clever. Unlike Patricia, Elizabeth rarely asked others for help and tended to work independently. In the lesson we taped on her she was getting bored with the work set and couldn't work out what to do:

. . . still came out wrong. I'm tired of this. I'm not doing it any more if I get it one more time wrong . . .

Immediately after this the teacher approaches and singles her out for praise:

Children look at this neat and very good work. I want work like that . . . I want work at that standard.

Without trying, Elizabeth has been able to attract the teacher's attention and at a particularly crucial time it stopped her from becoming disheartened (she continued the task) and reaffirmed her position as a good child who did her work in the 'correct' way. This made her powerful in the social relations of the classroom. All the others considered her good at her work and she was sought out for help even though she only gave it unwillingly.

Thus, the practices of the classroom situate each of these girls in quite different positions, Patricia as powerless and Elizabeth as powerful. Moreover, her positioning helps Elizabeth to deal with problems. Even when she does not get her work right, her problems are dealt with in an opposite way to those of Patricia. She is treated by the teacher as able, and therefore knows that if she tries she will succeed.

The helping/sub teacher positions occurred in both primary schools J1 and J2, but were the results of different practices used to teach mathematics. At school J2 the children were organized into groups to do specific pieces of work while at J1 the whole class most of the time worked on tasks together. At J2 the children were organized into groups to talk to and help each other and specific children were told to help others. For example, Sally was told to help Mollie:

> T: Now Sally, . . . that page is a page that you did very well. Do you remember organizing that? I'll give you a chance to organize with her, all right?

The teacher returns later to make sure all is well:

> So you understand what you're doing? Do you? You don't. Well, Sally, I thought you'd explained it to her. All right, you explain it again and I'll sit in on what you're saying.

Later, the teacher continues to discuss the question which is about sets, but as the conversation continues her exchanges, although nominally with Molly, are actually with Sally.

> T: . . . So here you are, you had twenty-four rows of bushes. I want your help (to Sally). You've got twenty-four rows of bushes and you set them out . . . how? Is that how you set them out? Is she copying this example or not?
>
> S: No Miss, she's not.

One of the examples from this school involves two boys helping each other. This next exchange whilst partly humourous is indicative of how George approached his task of helping Ray. The task is to find the area of three pieces of carpet. First they have to estimate it, then measure it, and compare the two sets of figures.

> G: Right, just tell me what length is?
> R: From there to there (touches top two corners)
> G: From there to there (top to bottom corners of one side)
> R: Correct (Stewart pretends to 'nut' him). What's area?
> G: Area? The whole.
> R: Right . . . you're learning.

It also expresses how Ray also accepted the parameters being used both to define himself and George and to define the way of working. We include this example to show that these positions are not *essential possessions* of boys or girls. They may relate to the production of femininity and masculinity which means they have different *effects* when displayed in the two sexes. However, it is important to show that they can, and indeed do, overlap.

Continuities and discontinuities

Let us look now at these same children in their first year of secondary school. In the chapter on teacher interviews and in the opening chapter we pointed to the differences between the sectors of the educational system. Essentially, apart from size, the main difference is in the backgrounds and framing of the staff, most of whom are subject specialists with different views on teaching and learning. As it turned out most of our sample had two mathematics teachers splitting three periods a week in a 2:1 ratio. Thus another dimension was introduced, the differences between teachers in their readings of the children's performances.

Patricia's position at secondary school did not change much. Both teachers (Mr G and Mr H) considered her weak and one repeated the prescription from primary school that he saw her as part of a group considered 'much of a muchness'. She still had difficulty attracting attention and was often dispossessed of her place in the queue by more aggressive or dominant children. The taped lesson was about learning braille and attempting mathematically to discover the combinations of dots possible. By the end of the first tape the mathematics teacher came to see Patricia and again explained the task to her, to which her only response was 'Yeah'. As in the primary school she was constantly asking her friends what to do: 'Can you do that one?' 'Tell me what to do.' The teacher saw her less than any of the others and when he did so it was for a shorter period of time. She rarely questioned him and rarely seemed to have got what she wanted from the exchange. Mostly she sat quietly — one of the teachers called her 'non-noticeable' — her only discussions with the others being about her small nephew or to ask for help.

For George (from School J2) the pattern had changed, and his position was reversed. At this stage he was the one in need of help. SMILE work as operated by his teacher meant mostly individual work. George's only sustained contact throughout the lesson taped was with another boy, John. John's sub-teacher relationship to George had developed as a result of the way in which the work was structured. George was constantly questioning him:

What page did you start from? . . . Have you done these? . . . What's that one?

Throughout the tape John helps George but in a way which affirms his superiority and George's inferiority:

. . . you're gonna have to do it, you're that stupid.

This relationship, between the two boys, shows clearly the assymetricality of the sub-teacher/helper relationship. George's position as the *helper* in the primary school derived from the teacher's power in setting up the relationship; as did his position as the *helped* in the secondary school.

At secondary school Elizabeth was considered: 'obviously fairly competent' by one of the teachers and by the other as '. . . confident . . . keen and eager. She wants to learn and she knows it takes some work on her part.' Elizabeth finds it easy to command attention from both her teachers. The rules of the classroom run by Ms R were fairly explicit in that children were praised for concentration — and told off for lack of it. They were also praised for good work, although the teacher often said that she didn't care about neatness but was more concerned with understanding what was written in books. Idle chat and gossip were frowned on, as was aggression. The girls were often told off for being trivial.

Elizabeth appeared to have none of these faults, although she was often aggressive to her neighbours if they could not follow her explanations or were being silly. She was easily discouraged when work was not going well (as we saw when she was at primary school) and claimed to be bored by the easiness of the work.

Ms R told the children that what they produced was the measure of what they could do and Elizabeth was often 'fed up' if she couldn't perform as she wished. In Mr P's lessons Elizabeth more often claimed to be bored — especially during SMILE lessons. Her confidence was manifest on the occasions observed when she was able to approach the teacher and demand a specific card be put on her matrix. Again, she helped others with their problems and was sought out for this purpose, even though she was often not very welcoming and could be very rude. She seemed to prefer Ms R's lessons, taking a dominant role, answering and asking questions and putting her head down and seeming to concentrate hard on the matter in hand.

If Elizabeth needed help she was able to obtain it almost instantaneously by demanding attention, and she often adopted the strategy (most favoured by boys) of just calling out answers. Elizabeth acted in what could be termed a 'masculine' way in that she was active, even aggressive, participating in all the classroom interactions. She was confident, especially about mathematics: 'I like it, I enjoy it. I can do it.' Although stylistically she acted in ways which were most often used by boys she did not want to be like a boy, nor did she like boys, choosing them in both primary and secondary schol as people she most disliked. A masculine positioning with respect to academic practice does not, therefore, mean a wholesale or unitary masculinity. Her success had to lie precisely in managing a 'balancing act' of a 'masculine' placing in relation to academic work and feminine one in respect of helpfulness and non-work contexts, which ensured a stable feminine position. This is neither necessarily easy nor successful for many children.

Following and challenging rules

Let us look now at one of the areas we discussed earlier — the problem of following or breaking the rules. As we have said, it was considered important to the learning of mathematics to be able to 'break set', to 'free' oneself from the confines of simply following rules or learning by rote in order to discover for oneself. Our teacher interviews — both primary and secondary — are full of the distinction between mechanical and creative work, between flair, or natural ability, and hard work. The following examples will serve to show how certain types of behaviour were 'read' by teachers as exemplifying these dichotomies.

In the lesson taken by Mr H in the secondary school about codes and specifically braille, the differences between Patricia's position and that of another girl Charlotte exemplifies some of the points made.

He begins by asking them how many combinations of dots they could find, given that they could use any combination of up to six dots. Charlotte volunteers an answer and so does Jim, his answer depending on leaving one block blank with no dots at all. The teacher agrees that this is a possibility and is immediately challenged by Charlotte: 'You said to count the dots'. The teacher obviously doesn't want to get into too much of a discussion and consequently uses the device of shifting mathematics out of the 'real' world into a plane of its own by replying: '. . . I think it's just as good maths if you count this one or you don't as long as you made it clear what you're doing'. So, 'in maths' anything goes as long as there's a reason for it. Charlotte, however, does not let things slip: twice she says: '. . . we were talking about dots'. She is implicitly questioning his definition and insisting on her own. The teacher ends the discussion — which he can do, holding as he does the ultimate authority — by moving on to something else, pretending that her query is a non-problem: '. . . I'm quite happy for you to take that attitude. I'm not arguing with you . . .' Being in charge and setting the parameters for the lesson in that he is the only one who knows what is going to happen, he moves on. The teacher again explains the task: that braille is for blind people and consists of raised dots on a rectangle. Since the teacher began to discuss the task Patricia has kept quiet.

Charlotte is able to challenge the teacher and attempt to assert her own definitions of what counts in the lessons. She is very demanding of the teacher's attention, using a variety of tricks and strategies to obtain it. For example, she never queues, always by-passing the others and seeing the teacher immediately. Patricia on the other hand queues patiently for long stretches and is often passed over two or three times in favour of other, more aggressive children — often girls. Charlotte is very confident in class lessons calling out answers, questioning and challenging the authority of the teacher. On two separate occasions she was observed berating her teacher about SMILE, which she did not consider 'proper' mathematics. Not all of the attention she receives is positive — far from it, she is constantly told off for talking and not working hard enough. But this indicates that the teacher is constantly monitoring her progress. It is interesting to examine what the two teachers of this class, Mr G and Mr H, have to say about Charlotte and Patricia and how that fits with a notion of following and challenging rules as essential to the 'correct' learning of mathematics.

It is Charlotte who generates the most interest. Charlotte, for Mr G 'is the one with the keenest brain in the sense of ideas. And she's the great problem solver.' Charlotte is discontented with SMILE which she has told her teachers she thinks makes them lazy '. . . because it's an easily manageable course . . . all you do is come in and the kids mark their own work . . .' The idea of the children having to do all the work she finds wrong: 'She's constantly trying out ideas all the time and that's why she finds SMILE a bit of a constraint.'

He recognizes her faults; selfishness, being mercenary in her use of her friends, and yet '. . . she's a tremendous abstract thinker, she's great at the maths. That, perhaps, we don't recognize enough.' He considers he might be at fault in not using the matrices imaginatively enough.

Overall Mr G believes in 'natural' ability and flair but is unable to be more specific. Nor is it clear whether he considers the cleverest children those who are most noticeable or those who cause him no trouble vis-à-vis his teaching technique.

Mr H felt that there were three children who 'really stood out' as good, two boys and a girl. He, too, felt that the good children were not of a piece and were not a recognizable group. He felt the girl to be 'a really outstanding mathematician', 'quite a phenomenon' who could do anything she was given. The boys were seen as confident and flexible. Another good girl he described as 'sharp . . . and quick at analysing a situation'. Finally he was able to sum up his feelings about all four of them that 'They have a certain sort of confidence.' It was this confidence which distinguished them '. . . they're not afraid of getting something wrong . . . They're prepared to try an answer, because if it's wrong then they'll just try the next one, or another approach.'

With regard to Charlotte Mr H saw her as being better verbally than her written work had led him to suppose. He was impressed with her work during the lesson on braille — surprised even since 'She hasn't done a great deal of work. She's always hard to get to do work . . . what made me think was that there is a sort of mathematical thinking going on in (her) which has never been tapped by me.'

What about Patricia? Both teachers, as we said earlier, saw her in the bottom four of the class. Mr G classed her, along with the other poor children, as '. . . just finding it difficult to understand things'. He thought that Patricia and her friends would concentrate and work through something, although it worried him that they wanted to '. . . get the matrix finished and carry on with another one' regardless of whether or not they had understood the work. He saw this as a particular problem for the poor children and a defect of the SMILE system. As a strategy for dealing with what seems to him to be a lack of confidence in the poor children Mr G sometimes set children like Patricia and her friends the same cards or their matrices: '. . . in many ways to give them a little bit of security if they're sharing the same problem'. This tactic also expresses a wider concern on this teacher's part that the three girls, Patricia, Jane and Sue, are in danger of becoming disaffected with school for the reason which he sees as: '. . . the big frustration thing. Not being able to do the tasks . . . I'm not trying to stretch any of them . . . I'm just trying to make them feel confident.'

Mr H simply sees her as '. . . a very non-noticeable pupil . . . She sits very quietly, on the whole getting on with things, not asking very much. I think she's not confident . . . she's slow really . . . on the whole doesn't search me out at all.' Of course Patricia does try to 'search him out' but is often discouraged in her attempts and therefore withdraws, whereas Charlotte by her confident persistence and refusal to be discouraged attracts attention without any problem. On the other hand, Ray is at least as disruptive as Charlotte, yet his 'rule challenging' is not sanctioned and is considered disreputable and certainly not seen as evidence of 'correct' learning. He is always in the middle of noise and disruption and disturbs other children a lot. Unlike Charlotte who is offered various inducements to work (like different work cards, to keep her interested) Ray is simply admonished and told to get on. His form of challenging, shouting, laughing, etc., is less acceptable than Charlotte's whose challenges to the teacher's authority are on an intellectual level.

One more example from a different class will suffice here. We shall compare Helen, one of our sample children, with another girl in her class, Kay, considered by her teachers, Messrs. J and K, a good girl.

Mr J said: '(Kay) registers most with me . . . because she does have a very precise interest. She likes to get things straight, she's almost officious about it.' He thought she was one of the few children able to make 'connections between the different areas of their work'. Mr K thought her an ideal pupil in that she could 'combine presentation with a thorough understanding of the mathematical ideas and in some ways that's an ideal combination'.

It was in the mention of presentation that we see a clear dichotomy between rule challenging and rule following which also counterposes girls and boys. For example Mr K's best children were able '. . . to organize . . . write down their work systematically . . . so that they can make sense of what follows'. However, later he categorizes his poor girls, of whom Helen is one, as having a 'tendency to over-accentuate their presentation' in which he feels 'all that kind of detailed work . . . hides the main mathematical concepts'.

So there is an immediate contradiction between the importance of presentation and challenging the rules, even though the former is considered an important part of mathematics. A good boy did not '. . . bother about his presentation, but I can see that he understands the mathematical ideas' and so Mr K chooses (as he himself says later) not to complain about the untidiness of his work in case it hinders his progress. This teacher's stress on presentation is unacknowledged and thus creates a double message and also a double bind for those who wish to do what he wants. Helen was considered not only to over-accentuate her presentation but also to be '. . . very well mannered and polite and rather than push herself forward to understand more mathematics she'll sit back . . .' When asked, Helen had this to say about mathematics: (They) try to explain . . . but there's always someone round . . . waiting and I feel I've got to hurry and I get in a mess.' Helen's diffidence and uncertainty caused her problems and she was the only person interviewed who mentioned being bullied. Other children in her class also mentioned that she was bullied. Kay, on the other hand, was seen very positively by all the children. She was the only first-year child out of the thirty-two interviewed who was successfully able to bridge the sex barrier and be accepted by boys and girls — on their terms. She was able to partake in masculine activities, particularly football, and still be seen as feminine as this quote from Ruth shows: 'She enjoys a game of football and all this as you see but the thing is . . . she don't wear trousers. She's just like a boy but she don't wear trousers. I'd like to be like that. She's fun.'

She was very powerful in the class, liaising between girls and boys, and teacher and pupils. Kay was mentioned by the boys as well in the same terms they would use to describe their own male friends. Here is Julie for example: 'Kay, she doesn't grass if you do something wrong. She's a good friend, she sticks with you . . . She's not stupid.' Willy likes Kay for more pragmatic reasons: '(She) supports the same football team as me . . . She's good at fooball . . . she's a brilliant defender . . . if the ball goes past her she'll bring you down.' Helen was more acerbic claiming that although the basis of Willy and Kay's friendship might be to do with football '. . . she goes round with him for things she doesn't know'. Kay is mentioned by both teachers as a visible child who is strong minded and insists on what she wants from both teachers and pupils.

So Kay is much more powerful than any of the other girls so far because she can place herself in relation to the teacher, the other girls and, unusually, the boys without any loss of 'femininity' or being considered 'odd'. This brings us on to our final point which is to attempt to synthesize all the data which we have collected.

Throughout our argument has been that attainment in mathematics is much more complicated than an ability/performance model would suggest. We have suggested that attainment in itself (or lack of it) is not a unitary possession of individual children. We have shown how teachers conceive of ability or attainment in ways which relate to the ideas they have about teaching and learning as indicative of certain processes. We have shown how social relations within the classroom serve to build up positions for children. In this way children are constructed as good or poor at mathematics and this has material effects on the way in which they see themselves in relation both to their own work and the other children.

We have theorized this in terms of a nexus of positions which children fill. In this way we can see both similarities between children and the specificity of each case, not as exceptions which can only be explained as such but rather as explicable in terms of the network of positions created in the classroom. In terms of this network of positions let us look at two of the girls we have already mentioned several times.

Patricia's position, for example, in the primary school is one of invisibility. With regard to the teacher she was considered not good enough and was rebuffed when she was finally able to make contact. His view of her in turn trapped her into reliance on her friends. She was revealed as anxious about her work. The other children saw her as in need of help which they provided whilst all the time cementing their positions as better than she. She was in a powerless position, on the periphery of the class into which she was unable to fit.

Elizabeth on the other hand filled the 'ideal' position at primary school in that she fitted squarely into the framework used by the teacher to understand good performance, i.e. her work was neat, she answered questions correctly but did not constantly need help or reassurance. In turn this gave her confidence in her own ability vis-à-vis both her work and her peers.

It is interesting that in the secondary school both these children replicated the positions they occupied at primary school and for the same reasons; Elizabeth fitted within the framework of relations in the classroom and Patricia did not.

Helen on the other hand was in a very different set of positions in the secondary school. At her primary school she was considered good at her work (contradicting her secondary teacher who thought she was poor) and although she was often forced, unlike Elizabeth, to seek help from the teacher and others she had few problems coping with the work. It is interesting that in her case study at primary school, a prediction was made about her transfer:

> Helen might find the transfer to secondary school more difficult because of her reliance on the definitions of others than, for example, Elizabeth who is much more self-contained.

And so it turned out. From a position of strength and relative confidence Helen moved to become more diffident, less sure of herself. She was being bullied and was no longer assured of approval from her teachers. This had left her adrift and unsure of how to slot into the positions of her class.

Fourth-year secondary

By the fourth year the children were established within their classes both in their relationship to other children and to the subject. Throughout the school the emphasis on individual work and individual credit meant that by the fourth year the children had become totally individualistic about their work. As we showed in the last chapter, the older the children were the less part of the school group they felt. We could surmise that working on individualized mathematics schemes for so long left them more concerned to get on than to discuss and/or to help others. Also, it seems to be the case that by the fourth year all the children were doing such different things that there was little common ground for discussion about work. Let us look at the development of the categories of analysis we have used so far.

It seems significant that the girls seen as good by the fourth-year teacher rarely sought help with their work. In this way it was possible for the teacher to see them as good, although hard working. They both helped other pupils and were seen by them as good. They are both described by the teacher as 'conscientious' even though one of them had been truanting.

For the poor girls Yong was considered to work hard but not to be very good. She rarely went to the teacher for help, though, through a combination of timidity and her own feelings that she was not very good, finding mathematics a struggle. Again, as in both the classes lower down the school, it seemed to be the ability to attract the teacher's attention which differentiated the good from the poor pupils. Yong in her interview expressed her feelings as follows:

I'm not very good at maths . . . I haven't talked to him much about maths. I haven't asked him much . . . Sometimes when I'm not good at maths I don't ask.

'I always struggle', she said of herself. She felt trapped, afraid to ask yet needing help. This seemed to give her a feeling of powerlessness in her mathematics lessons and she was forced to turn to others for help and support. In the video-taped lesson she spent a lot of time (nearly twenty minutes) reading and re-reading her work card. She tried to work back from the answers in the answer book. At this point the teacher came beside her:

T: Let me see . . . you might, where you've got a difficult graph like this, you might find it better on proper graph paper. Do you know what I mean?

Y: Yeah. *(The teacher moves away leaving Yong rubbing out her work)* I was all wrong. *(She tries again and attracts attention of a visitor to the classroom who goes to help)* I don't get this *(reads from work card)* You are driving 200 ft behind Bruce. It says 'what is your maximum safe speed?'

V: Right, so you just read that off the graph . . . I've not actually seen this card before. Do you read off the graph? The maximum safe speed is 55 it looks like, isn't it?

Y: Oh yeah . . . D'you get what this means? Do you think the curve should go through the nought?

V: Your curve does actually go through the nought . . .

Y: Oh.

V: Think about what nought means. If you're no distance at all behind the car in front, that means you're actually touching the car in front . . . your speed actually has to be nought for it to be safe, doesn't it?

Y: Mmm.

This extract expresses some of the dilemmas mentioned earlier that were experienced by both teachers and pupils. The pupil had no access to, or was afraid to express, the terms in which the activity was conducted; the teacher wished neither to push nor to do all the work for the girl, preferring to let her work out the answers for herself. This trapped both participants and, in this case, contributed to the girl's own very pessimistic view of her own ability.

Good children, regardless of sex, helped other children. Michelle asked to help a boy claimed, however: 'I've forgotten how I did it.' When the teacher came to ask what the problem was she replied: 'I can't do it anyway. I doubt if I'll be able to understand it. I don't understand any of this.' This girl was at the intersection of the constellations of being clever and helpful and seemed to want to abdicate some of the responsibility which this appeared to imply.

As noted earlier, it was most important for these pupils to be seen to be able to handle social relations properly. Helping/being helped, therefore, took on other connotations to do with being 'boring', 'unable to do it' and therefore not an interesting person. It was as 'uncool' to help as to be helped unless it was in a casual way as in this sequence between Viv and Michelle:

V: I don't get that.

M: *(Who has been sitting on top of a store cupboard)* What?'

V: 'x' equals two.

M: 'x' 've you gotta do it on? Where? Where does it say that? . . . Is 'x' going along the bottom?

V: So I put 'a' on two.

M: 'x' equals two.

V: I don't get that. First of all thought it mean a *(untrans.)* axis and put 'a' here, along the 'x' axis.

M: No . . . that's wrong, that's two, that's . . . It's 'x' equals two, that line there.

V: Oh God, I always get this wrong. 'x' equals two would be here.'

M: Yeah.

V: Oh I get it now, thank you.

By this stage it seems that a lot of the children have decided mathematics is 'boring'. Few seem able to be very articulate as to why they find it so, apart from blaming themselves. We could suggest that it seems unreal. Michelle in this last extract seemed to understand her friend's problem, but makes no attempt to explain it. Equally her friend seemed content to assume that it was her fault. ('I always get this wrong.') At this stage power is invested in those who have access to mathematics as a body of knowledge and they retain their power by not letting others into their secret. Those who are good have understood the rule-governed nature of the interaction and are able to work within it successfully. There was much discussion about the amount of work done and this was related to examinations since a school based CSE Mode Three course derives 50 per cent of marks from course work for which a minimum number of matrices need to be completed. Rule challenging or following in the sense that this has been discussed earlier seemed at this level to apply more to those who challenged behavioural rules. This was not necessarily sanctioned (Michelle often truanted) but we saw little evidence of an ability to 'break set' even amongst the good children.

There was least discussion of mathematics amongst this age range. Work was done silently. Help tended to be offered practically, with little or no discussion of issues. Viv summed up a dilemma expressed by all the poor girls interviewed:

V: When I came to this school I had a Triple One which you get in primary school *(refers to band one on ILEA's comparability tests given in the primary school to allocate children to ability band one, two or three, prior to entry to secondary school. This is used as a way of making sure the secondary school's intake is properly mixed in terms of ability).*

I: Did you enjoy maths in your primary school?

V: Absolutely detested it.

I: What happened once you got here then?

V: I don't know, at some point I thought, ugh, I'm no good at maths and I didn't really try I suppose . . . now I'm trying to work again. But also I get easily distracted by people . . . who're good at maths, anyway so they can distract us and get back on with their work again.

However it would be unfair to suggest that the girls accepted the teacher's definitions of their ability unequivocally. As we have stressed throughout, the depiction of girls as passive is one of the areas to which we take exception in some of the work on gender-stereotyping. There were a variety of ways in which the girls in our sample resisted the teacher. Power relations within the classroom were constantly being renegotiated. Yong, for example, insisted, despite the teacher's experienced misgivings, in attending a group which was doing work more specifically aimed at the O level syllabus. Often, resistance took the shape of refusing to accept the teacher's definitions of what was considered appropriate and thus refusing to ask him for help, or truanting (see Hayward, 1983).

The videotapes of the boys in the sample show more interaction between them and the teacher about mathematics, with the teacher correcting work and discussing problems and answers with them in more depth. The boys were monitored more closely in relation to their work rate.

T: . . . Trevor you're not doing as much as usual. Mark's obviously a
bad influence.

It is the good boys — as chosen by the teacher — who seem to have most
contact with him about work. Yet, as we discovered, only one child was
entered for O level and that was a girl who insisted on the entry against
the wishes of the teacher. We were not able to explore this in more detail
because it took place after the period of our fieldwork. Like the girls, the
boys discussed amongst themselves their lives outside lessons whilst doing
their mathematics. All the children were doing different things so there
was little common grounds for discussion, although most of the children
would have covered most of the areas at some time.

The teacher was understood as a facilitator and not often approached
for help, although he carefully monitored the class and tried to see every
child as often as possible. Yet he, too, was caught in a double bind. He
worked to facilitate the children's learning by providing the resources and
a suitable environment. Out of respect for their feelings and aware of the
fears mathematics can generate, he tried not to push or to be authoritarian.
This, in turn, left the quieter girls struggling and with feelings that they
could not do the work and that the teacher was unapproachable — the
last thing, the very opposite of, what he wanted to convey.

In part this double bind had to do with the imminence of public
examinations and the necessity — since the school offered the choice —
of selecting for 'O' level and CSE groupings, and thus irrevocably labelling
some children as more clever.

We shall discuss in the final chapter the implication of and conclusions
which can be drawn from this brief overview of our work.

Summary

In this chapter we have argued that relations of power and powerlessness
in the classroom are not fixed but constantly shifting. Girls can be successful in terms of mathematical attainment, gaining power by taking
responsibility in the classroom, but remain relatively powerless in terms
of teachers' judgements of their performance. Since the latter depend on
indications of the challenging of rules which are understood as 'real
understanding', 'flair' or 'brilliance', girls often are left in an ambiguous
position.

Our analysis of individual cases exemplified some positionings, occupied by a variety of girls (and boys). We focused on the dimensions of helping and being helped, suggesting that aspects of femininity, cohering around positions of being nice, kind and helpful, meant that some girls could be helpers and even the poor girls could at least succeed in being nice to others.

Some good girls have to manage a 'balancing act' between feminine and masculine positions, and the girls who are most academically successful *and* simultaneously popular with peers, achieve both. Such girls are visible in the practices of the classroom. The poor girls remain invisible, unnoticed. They do not like to ask for help (and the teachers like the girls to work out the answers for themselves). Girls can occupy positions which are related to masculinity, for example, challenging the procedural rules of mathematics. Yet, such positions are never unambiguous nor gender-neutral, because of the effects of other positionings on the girls in question.

By the fourth year of the secondary school girls are still in similar positions, although interaction with the teacher is far more restricted and mathematics has become a more individualized (and secret) activity. Many children by this time find mathematics boring.

Chapter Seven
Conclusions and Implications

Let us begin by summarizing what we take to be the central issues raised by our research.

1. The results of tests given to children in the top of the junior school, first year of the secondary school and the fourth year of the secondary school, suggest that, contrary to our initial hypothesis, there is no simple discontinuity of performance between ages 11 and 16. We had put forward such a hypothesis in order to account for the fact that girls' performance relative to boys in the primary school was certainly not poor. In order to account for what appeared from the statistics to be a performance decrement by the age of sixteen, we had argued in favour of a discontinuity of performance rather than a continuity of poor performance.

What is suggested by the test and other data, is that there is far more continuity than we had imagined. However, it is continuity of good, rather than poor performance. Even in the fourth-year secondary tests, girls are still doing well in comparison with boys. Such a finding suggests that what is necessary is an adequate and more complex explanation to account for both continuity and the final result of girls' relatively poor O level success when compared with the pass marks of the boys.

2. It is this more complex explanation to which we have addressed ourselves and summarized in the remaining chapters of this volume.

The variation of test performance from classroom to classroom was an important initial starting point in relation to possible explanations. We examined the relationship between the production of theories of teaching and learning, the ideas expressed by particular teachers, their classroom practices and the performance and identity of particular pupils understood within the framework of the social relations of specific classrooms.

3. Examining the relationship between ideas about the teaching and learning of mathematics and classroom practices produced a reading of success and failure in which actual attainment is no longer a simple or reliable indicator of success. The move towards 'real understanding' and away from 'rote memorization' means that certain characteristics are invested in the individual. It is, therefore, possible to be successful for the wrong reasons. That is, a child may do well, but may be suspected of 'not understanding'. Teachers are, therefore, at pains to promote such understanding in their practices and to look out and find, evaluate and remedy evidence of success for the wrong reasons. This means that their evaluations are made in terms of the presence or absence of these attributes.

Such practices have particular effects, in that the characteristics taken to be indicators of 'real understanding' are to a large extent co-terminous with those used to describe masculinity. There is, therefore, a problem for many girls: their success is double-edged; the characteristics of femininity which they display lead teachers to assume a lack of understanding. This has particular effects in terms of their classroom performance and in terms of the practical consequences of the disjunction between success and its evaluation.

4. The analysis of the data from the children's interviews, repertory grids and classroom activities suggests that girls struggle to achieve a femininity which possesses the characteristics which are the target of teachers' pejorative evaluations. We use the word 'struggle' advisedly. The data does not suggest an easy or natural process, but precisely a struggle to be understood/understand themselves in certain ways, which have particular effects on the social relations of the classroom.

Our argument is that girls are located at the nexus of a constellation of practices characterized by the tortuous and contradictory relationship between gender (masculinity/femininity) and intellectuality (academic performance and attainment). Femininity particularly appears for the girls to be related to such characteristics as 'helpful', 'nice', 'kind' and 'attractive'. These are precisely the characteristics which help to render them good and hard working in the classroom and so lead to academic success, but not to a display of those characteristics which are read as indicating 'real understanding', 'flair', 'brilliance' and so forth.

How and why the positionings in different practices cross over and work in such ways as to produce different effects in particular girls was not the object of this enquiry, but it is examined elsewhere (Walkerdine, in preparation). We believe it important to examine and produce in more detail an analysis which explores the how and why of particular individuals and their specific positioning. In other words we would argue that it is necessary but not sufficient as an explanation to understand the girls as positioned in relation to the normative and normalizing effects of the practices. We still need to understand the processes by and through which this occurs.

5. We have paid particular attention to the phenomenon described as rule following/challenging. We examined the way in which following the procedural rules of mathematics and the behavioural rules of the classroom was necessary to successful completion of tasks. However, challenging the internal rules of the mathematical discourse, relating particularly to the teacher's authority as guardian of those rules is important in producing what the teachers describe as 'real understanding'. Such challenging requires considerable confidence because it necessitates the recognition that rules are to be simultaneously followed and challenged. That many girls do not have such confidence, nor would dare to make a challenge offers a different explanation of girls' mathematical development than one which relies on the naturalistic and immutable. Such an approach as ours offers a profound challenge not only to practice but to theories of cognitive development and of masculinity and femininity.

6. Girls in our study are more positive than boys about mathematics right into the fourth year of the secondary school, but by this time many girls (and boys) had given up in a disheartened way because they find the subject boring. Their teacher was disheartened especially since a radical child-centred scheme was used supposedly to retain and generate interest. Such a relation cannot help but have effects on performance and attainment.

7. We have been critical of approaches to girls' performance which concentrate on the 'hidden curriculum', because they do not engage with the processes which overtly define knowledge and the regulation of educational practices. Our analysis suggests that such processes and the consequent production of positionings for children are crucial to an understanding of gender and schooling.

8. Individualized schemes of work such as SMILE help to protect the less confident girls, but at the same time offer no basis of practices which might produce the possibility of confident challenge.

Implications

We believe that these conclusions drawn from our research have profound implications for the issue of girls and mathematics, and for gender and education more generally. Here we shall sketch out some of these implications. They will be explored in detail in later publications.

We have already mentioned that we are arguing for a paradigm shift in the terms in which the issues of girls' performance in mathematics have been understood. We shall not labour the point here, except to note that we have only just begun to open up a way of looking at these issues in a field which deserves considerably more attention. In this short section we want to concentrate on certain issues of practice which should be raised in the light of our analysis.

Previous and present interventions in practice in relation to girls and mathematics in particular, and girls' education in general, have concentrated on a variety of strategies ranging from equality of opportunity (the assuring of equivalence of provision for boys and girls) to attempts to shift attitudes. Our analysis, in presenting the problem rather differently, is critical of such approaches. For example, those approaches which have sought to increase young girls' autonomy by the provision of missed play experience and the provision of construction toys, etc., assume that there is a problem of girls' incomplete development. We have argued that there are, indeed, problems for girls, but they are not simple problems of failure which a strategy such as this has been designed to remedy. We see the need for a dynamic model of learning in which children are not simple passive recipients of external social forces. Several existing approaches, unfortunately, operate as though this were the case. Attempts to introduce better role models for girls, to change stereotyped images in books and presentation of subject matter, to allow greater curriculum choice and provision, operate on a model of rational change through the provision of information. If we are correct in our analysis of the way in which girls struggle to achieve femininity, no amount of information will produce change by itself. Indeed, one is tempted to suggest that it might well provoke serious resistance. Often when approaches to change which use stereotyping fail, the automatic response is to argue that, after all, girls' femininity is 'natural' and therefore unchangeable. We have tried to demonstrate that other conclusions may be drawn from the failure of such approaches. This is why we suggest that practical and theoretical changes must go hand-in-hand.

Now clearly, while we have argued against a simple model of failure, many girls do fail. There remains, therefore, a pressing set of problems.

We would support interventions which attempt to work *with* the contradictory positions in which the girls (and their female teachers) are placed. We would support approaches which examine girls' own fears and feelings, as well as their performance. There are already some schools (particularly secondary schools) which try to work in this way.

Lastly, how can we now understand the situation with respect to girls' examination performance at sixteen? Since our research was conducted in inner London, let us take figures supplied by the Inner London Education Authority (1981, 1982) which are broken down by CSE and O level. Here we see exemplified the tendency to enter proportionately more girls for CSE than for O level. Additionally, if we examine sex differences in the O level and CSE passes, while they are clearly present, they are not large. In fact other studies (for example Murphy, 1980) indicate that girls achieve a generally higher pass rate than boys in all subjects in both O and A level. Such figures are clearly important in the light of our own work.

In other words there is here presented no massive fall off in performance. While it is difficult from these figures to make any claims about the whole country, we might expect variations which relate to regional and school policies on examination entry. O level, for example, is intended to be taken by only 25 per cent of sixteen year olds. It is, therefore, difficult to base arguments about girls as a category on an examination intended for a quarter of the total sixteen-year-old school population. However, even taking this into consideration, in the school which we studied the entry for O level was only 13.5 per cent. There were, as we have seen, particular reasons for this. The school favoured CSE because it was school based and therefore felt to be better for less confident pupils. Such an approach could well either be repeated or reversed in other schools. It is, however, premised on the view that CSE grade one is equivalent to an O level pass and, while this may be the case in theory, it is abundantly clear that this is not so in practice. Universities adopt an unfavourable attitude towards CSE, as do many employers. It could be argued, therefore, that one problem might be solved if more girls were entered for O level. Clearly, this may well help a *small minority* of girls. However, more to the point seems to be the existence of a two-tier examination structure, at sixteen. It seems to us that our data certainly support an argument for a common examination like that which is to be introduced in the summer of 1988. The remaining issue, however, is what kind of examination is appropriate.

There are certainly arguments in favour of developing a system along the lines of CSE Mode Three which allows teacher and school autonomy and therefore flexibility according to local conditions. Such an idea may raise fears about lowering of standards, but it can be pointed out that currently all universities set different curricula and have an autonomous examination structure. There are, however, arguments against such increase in school autonomy (for example Finn, Grant and Johnson, 1980). Our analysis certainly suggests that in the current system the power and effect of teacher evaluation and judgement of pupil suitability and performance is immense. There are grounds for concern, therefore, that increased provision along the lines of Mode Three would not increase girls' chances of success, despite the best intentions of teachers to promote the welfare of their less confident pupils.

It is both interesting and important to note that, just as with our pilot work we discovered the phenomenon, apparently well known and yet also hidden, that girls perform well in primary schools, the same phenomenon holds true for secondary education. In France girls are said to adapt better to secondary school, while in Britain and the USA girls have a higher level of attainment (generally speaking) than boys in secondary school as measured by examination passes. Is this once more to be 'explained away' by recourse to an argument about greater maturity? When, one might wonder cynically, can boys ever be said to have matured?

Such statements bring into sharp focus the close relationship between theories which explain performance, the performance itself and the practical effects for girls. This paper is able only to raise these highly contentious and debatable issues, but if we have succeeded in putting on the agenda central problems concerning the education of girls and women, our research will have been well worthwhile.

Bibliography

Assessment of Performance Unit (APU) (1980a), *Mathematical Development Primary Survey Report No.1,* January. London: HMSO.

_____ (1980b), *Mathematical Development, Secondary Survey Report No.1,* September. London: HMSO.

_____ (1981a), *Mathematical Development, Primary Survey Report No.2,* June. London: HMSO.

_____ (1981b), *Mathematical Development, Secondary Survey Report No.2,* December. London: HMSO.

_____ (1982a), *Mathematical Development, Primary Survey Report No.3,* May. London: HMSO.

_____ (1982b), *Mathematical Development, Secondary Survey Report No.3,* December. London: HMSO.

Atkins, L. and Jarrett, D. (1979), 'The significance of "significance tests" ', in I. Miles and J. Evans (eds.), *Demystifying Social Statistics.* London: Pluto.

Bannister, D., and Fransella, F. (1971), *Inquiring Man: The Psychology of Personal Constructs.* Harmondsworth: Penguin Books.

Barnes, D., Britton, J. and Rosen, H. (1969), *Language, the Learner and the School.* Harmondsworth: Penguin Books.

Central Advisory Council on Education (1967), *Children and Their Primary Schools* (Plowden Report). London: HMSO.

Clarricoates, K. (1978), 'Dinosaurs in the classroom: a re-examination of some aspects of the "hidden curriculum" in primary schools', *Women's Studies International Quarterly,* Vol.1, pp.353-64.

_____ (1983), 'Girl or boy — does it matter? Some aspects of the "hidden curriculum" and interaction in the classroom', *Primary Education Review,* No.17, Summer.

Cohen, J. (1979), *Statistical Power Analysis for the Behavioural Sciences.* London: Academic Press.

Corran, G. and Walkerdine, V. (1981), *The Practice of Reason,* Vol.1, *Reading the Signs.* Mimeo, University of London Institute of Education.

Department of Education and Science (1982a), *Mathematics Counts: Report of the Committee of Inquiry into the Teaching of Mathematics in Schools* (Cockcroft Report). London: HMSO.

_____ (1982b), *Educational Statistics for the United Kingdom,* Attainments of School Leavers, 1979-80. London: HMSO.

_____ (1983), *Educational Statistics for the United Kingdom,* Attainments of School Leavers, 1980-81. London: HMSO.

Eddowes, M. (1983), *Humble Pi: the Mathematics Education of Girls.* London: Longman for the Schools Council.

Equal Opportunities Commission (1981), *Education of Girls: A Statistical Analysis.* Statistics Unit, EOC.

Eynard (Walden), R. and Walkerdine, V. (1981), *The Practice of Reason,* Vol.2, *Girls and Mathematics.* Mimeo, University of London Institute of Education.

Finn, D., Grant, N., and Johnson, R. (1980), 'Social democracy, education — the crisis', in Centre for Contemporary Cultural Studies (ed.), *On Ideology.* London: Hutchinson.

Galton, M. and Simon, B. (eds.) (1980), *Progress and Performance in the Primary Classroom.* London: Routledge and Kegan Paul.

Galton, M. Simon, B. and Croll, B. (1980), *Inside the Primary Classroom,* First Report of the ORACLE Project. London: Routledge and Kegan Paul.

Galton, M. and Willcocks, J. (1983), *Moving from the Primary Classroom.* London: Routledge and Kegan Paul.

Gibbons, R. (1975), 'An account of SMILE', *Mathematics in School,* Vol.4, No.6, November, pp.14-16.

Gilligan, C. (1982), *In a Different Voice: Psychological Theory and Women's Development.* Cambridge, Mass: Harvard University Press.

_____ (1983), 'Female development in adolescence: implications for theory'. Paper presented at the seventh biennial meeting of the International Society for the Study of Behavioural Development, Munich.

Goldstein, R. (1973), 'Learning to SMILE' ILEA *Contact,* Vol.2, No.21, 7 December, pp.17-19.

Gubb, J. (1983), *GAMA Newsletter,* No.4.

Hannam, C., Smyth, P. and Stephenson, N. (1971), *Young Teachers and Reluctant Learners.* Harmondsworth: Penguin Books.

Hartley, D. (1980), 'Sex differences in the infant school: definitions and theories', *British Journal of Sociology of Education,* Vol.1, No.1, pp.93-105.

Hayward, M.S. (1983), 'Girls, resistance and schooling'. Unpublished MA dissertation, University of London Institute of Education.

Henriques, J. et al. (1984), *Changing the Subject: Psychology, Social Regulation and Subjectivity.* London: Methuen.

Howson, G. (1982), review of *Girls and Mathematics: the early years, Education,* Vol.159, No.11, p.188.

Inner London Education Authority (1981), *Achievement in Schools II: Sex Differences.* Research and Statistics Report, RS 806/81.

_____ (1982), *Sex Differences and Achievement.* Research and Statistics Report, RS 823/82.

_____ (1983), *Anti-sexist Initiatives in ILEA Schools,* Research and Statistics Report, RS 868/83.

Johnson, R.C. (1963), 'Similarity in IQ of separated identical twins as related to the length of time spent in the same environment', *Child Development,* Vol.34, pp.745-9.

Kelly, A.V. (1978), *Mixed-Ability Grouping: Theory and Practice.* London: Harper and Row.

Kuhn, T.S. (1962), *The Structure of Scientific Revolutions.* Chicago: Chicago University Press.

Langdon, N. (1976), 'Smiling in the classroom', *Mathematics in School,* Vol.5, No.5, November.

Maccoby, E.E., and Jacklin, C.N. (1975), *The Psychology of Sex Differences.* Stanford University Press.

McRobbie, A. (1978), *Jackie: An Ideology of Adolescent Femininity,* Working Papers in Cultural Studies SP 53. Birmingham: Centre for Contemporary Cultural Studies.

Morrison, D.E. and Henkel, R.E. (eds.) (1970), *The Significance Test Controversy.* Chicago: Aldine.

Murphy, R. (1980), 'Sex Difference in GCE examination entry statistics and success rates', *Educational Studies,* Vol.6, No.2, June.

Northam, J. (1983), 'Girls and boys in primary maths books', *Education 3-13,* Vol.10, No.1, pp.11-14.

Nunally, J.C. (1960), 'The place of statistics in psychology', *Educational and Psychological Measurement,* Vol.20, pp.641-50.

Pintner, R. and Forlano, G. (1933), 'The influence of month of birth on intelligence quotients', *Journal of Educational Psychology,* Vol.24, pp.561-84.

Postman, N. and Weingartner, C. (1971), *Teaching as a Subversive Activity.* Harmondsworth: Penguin Books.

Salmon, P. (1976), 'Grid measures in child subjects', in P. Salter (ed.), *The Measurement of Intrapersonal Space by Grid Technique.* London: Wiley.

Sheffield City Polytechnic (1983), *Mathematics Education and Girls,* report of the research project. Department of Mathematics, Statistics and Operational Research, Sheffield City Polytechnic.

Shuard, H. (1981), 'Mathematics and the ten-year-old girl', *The Times Educational Supplement,* 27 March.

—— (1982), 'Differences in mathematical performance between girls and boys', in Department of Education and Science (1982a).

—— (1983), 'The relative attainment of girls and boys in mathematics in the primary years', *GAMA Newsletter,* September (Paper presented at the Girls and Mathematics Association (GAMA) Conference, 27 May).

Simon, B. and Willocks, J. (ed.) (1981), *Research and Practice in the Primary Classroom.* London: Routledge and Kegan Paul.

SMILE Centre (n.d.), *SMILE Guide.* ILEA Resources Centre, Middle Row School.

Spender, D. (1982), *Invisible Women: The Schooling Scandal.* London: Writers and Readers Publishing Co-operative.

Spender, D. and Sarah, E. (1980), *Learning to Lose: Sexism and Education.* London: Women's Press.

Walden, R. and Walkerdine, V. (1982), *Girls and Mathematics: The Early Years,* Bedford Way Papers 8, University of London Institute of Education.

_____ (1983), 'A for effort, E for ability: the case of girls and mathematics', *Primary Education Review,* No.17, Summer.

Walkerdine, V. (1981), 'Sex, power and pedagogy', *Screen Education,* No.38, Spring.

_____ (1982), 'From context to text: a psychosemiotic approach to abstract thought', in M. Beveridge (ed.) *Children's Thinking Through Language.* London: Edward Arnold.

_____ (1983a), 'Girls and mathematics: a reflection on theories of cognitive development'. Paper presented to the International Society for the Study of Behavioural Development, Munich.

_____ (1983b), 'She's a good little worker: femininity in the early mathematics classroom', *GAMA Newsletter.*

_____ (1983c), 'It's only natural: rethinking child-centred pedagogy' in A.M. Wolpe and J. Donald (eds.), *Is There Anyone Here from Education?* London: Pluto.

_____ (1984), 'Cognitive development and the child-centred pedagogy' in Henriques et al.

_____ (in press), *The Mastery of Reason.* London: Methuen.

Walkerdine, V. and Walden, R. (1981), 'Inferior attainment?' *The Times Educational Seupplement,* 3 July.

_____ (1982), 'We make girls miss out on maths', *Child Education,* Vol.59, No.9, September.

Walkerdine, V., Walden, R. and Owen, C. (1982), 'Some methodological issues in the interpretation of data relating to girls' performance in mathematics'. Paper presented at the British Psychological Society Education Conference, University of Durham.

Ward, M. (1979), *Mathematics and the Ten-year-old:* Schools Council Working Paper 61, London: Evans Bros. and Methuen.

Warnock, M. (1982), 'Last word — for now'. *The Times Educational Supplement,* 2 July.

Weiner, G. (1980), 'Sex differences in mathematical performance: a review of research and possible action' in R. Deem (ed.), *Schooling for Women's Work.* London: Routledge and Kegan Paul.

Whyte, J. (1983), 'How girls learn to be losers', *Primary Education Review.* No.17, Summer.

Willis, P. (1979), *Learning to Labour.* London: Saxon House.

Yates, A. and Pidgeon, D. (1957), *Admission to Grammar Schools.* London: Newnes.

Young, M.F.D. (1970), *Knowledge and Control.* London: Collier Macmillan.